Introduction to Chemistry

Introduction to Chemistry

Benjamin Schultz

WILLFORD PRESS

www.willfordpress.com

Published by Willford Press,
118-35 Queens Blvd., Suite 400,
Forest Hills, NY 11375, USA

ISBN: 978-1-64728-000-0

Cataloging-in-Publication Data

Introduction to chemistry / Benjamin Schultz.
    p. cm.
Includes bibliographical references and index.
ISBN 978-1-64728-000-0
1. Chemistry. 2. Physical sciences. I. Schultz, Benjamin.
QD31.3 .I58 2022
540--dc23

For information on all Willford Press publications
visit our website at www.willfordpress.com

WILLFORD PRESS

Table of Contents

Preface

The purpose of this book is to help students understand the fundamental concepts of this discipline. It is designed to motivate students to learn and prosper. I am grateful for the support of my colleagues. I would also like to acknowledge the encouragement of my family.

Chemistry is a discipline of science that deals with the elements and compounds made up of atoms, molecules and ions. It is also concerned with their composition, structure, behavior and properties, as well as the changes they undergo during a reaction with other substances. There are three major branches of chemistry, namely, inorganic chemistry, organic chemistry and physical chemistry. Inorganic chemistry studies the properties and reactions of inorganic compounds. Organic chemistry focuses on the structure, properties and mechanisms of compounds which have a carbon-hydrogen bond. Physical chemistry deals with the study of the fundamental and physical basis of chemical systems and processes. This textbook attempts to understand the multiple branches that fall under the discipline of chemistry and how such concepts have practical applications. It aims to shed light on some of the unexplored aspects of this field. Those in search of information to further their knowledge will be greatly assisted by this book.

A foreword for all the chapters is provided below:

Chapter – Introduction

Chemistry is a fundamental science that studies composition, structure and properties of compounds and elements composed of atoms, molecules and ions. It also studies the changes they undergo during a chemical reaction with other substances. This is an introductory chapter which will introduce briefly all the significant aspects of chemistry.

Chapter – Branches of Chemistry

There are numerous sub-disciplines within chemistry such as electrochemistry, physical chemistry, analytical chemistry, chemical kinetics, radiochemistry and solid-state chemistry. The topics elaborated in this chapter will help in gaining a better perspective about these branches of chemistry.

Chapter – Organic Chemistry

The sub-discipline of chemistry that studies the properties, structures and reactions of compounds that contain carbon in covalent bonding is known as organic chemistry. Some of the major areas of study within this field are functional groups and aliphatic compounds. All the diverse areas of organic chemistry have been carefully analyzed in this chapter.

Chapter – Inorganic Chemistry

The chemical compounds that lack carbon- hydrogen bonds are known as inorganic compounds. Inorganic chemistry focuses on the behavior and synthesis of inorganic and organometallic compounds. The topics elaborated in this chapter will help in gaining a better perspective about inorganic chemistry.

Chapter – Acids and Bases

Acid is a molecule that is capable of donating a proton and forming a covalent bond with a pair of electrons. Bases are the substances that release hydroxide ions in an aqueous solution. This chapter has been carefully written to provide an easy understanding of acids and bases as well as acid-base reactions.

Chapter – Chemical Mixtures and Solutions

Chemical mixture is a material that is made up of two or more different substances which are physically combined. The mixture could homogeneous or heterogeneous in nature. Solutions are a type of homogenous mixture which consists of a solute that is dissolved in another substance, known as a solvent. The diverse types of chemical mixtures and solutions have been thoroughly discussed in this chapter.

Benjamin Schultz

1
Introduction

Chemistry is a fundamental science that studies composition, structure and properties of compounds and elements composed of atoms, molecules and ions. It also studies the changes they undergo during a chemical reaction with other substances. This is an introductory chapter which will introduce briefly all the significant aspects of chemistry.

Chemistry

Chemistry is the study of matter, its properties, how and why substances combine or separate to form other substances, and how substances interact with energy.

Every material in existence is made up of matter - even our own bodies. Chemistry is involved in everything we do, from growing and cooking food to cleaning our homes and bodies to launching a space shuttle.

Branches of Chemistry

- Analytical chemistry uses qualitative and quantitative observation to identify and measure the physical and chemical properties of substances. In a sense, all chemistry is analytical.

- Physical chemistry combines chemistry with physics. Physical chemists study how matter and energy interact. Thermodynamics and quantum mechanics are two of the important branches of physical chemistry.

- Organic chemistry specifically studies compounds that contain the element carbon. Carbon has many unique properties that allow it to form complex chemical bonds and very large molecules. Organic chemistry is known as the "Chemistry of Life" because all of the molecules that make up living tissue have carbon as part of their makeup.

- Inorganic chemistry studies materials such as metals and gases that do not have carbon as part of their makeup.

- Biochemistry is the study of chemical processes that occur within living organisms.

Fields of Study

Within these broad categories are countless fields of study, many of which have important effects on our daily life. Chemists improve many products, from the food we eat and the clothing we wear to the materials with which we build our homes. Chemistry helps to protect our environment and searches for new sources of energy.

Food Chemistry

Food science deals with the three biological components of food — carbohydrates, lipids and proteins. Carbohydrates are sugars and starches, the chemical fuels needed for our cells to function. Lipids are fats and oils and are essential parts of cell membranes and to lubricate and cushion organs within the body. Because fats have 2.25 times the energy per gram than either carbohydrates or proteins, many people try to limit their intake to avoid becoming overweight. Proteins are complex molecules composed of from 100 to 500 or more amino acids that are chained together and folded into three-dimensional shapes necessary for the structure and function of every cell. Our bodies can synthesize some of the amino acids; however eight of them, the essential amino acids, must be taken in as part of our food. Food scientists are also concerned with the inorganic components of food such as its water content, minerals, vitamins and enzymes.

Food chemists improve the quality, safety, storage and taste of our food. Food chemists may work for private industry to develop new products or improve processing. They may also work for government agencies such as the Food and Drug Administration to inspect food products and handlers to protect us from contamination or harmful practices. Food chemists test products to supply information used for the nutrition labels or to determine how packaging and storage affects the safety and quality of the food. Flavorists work with chemicals to change the taste of food. Chemists may also work on other ways to improve sensory appeal, such as enhancing color, odor or texture.

Environmental Chemistry

Environmental chemists study how chemicals interact with the natural environment. Environmental chemistry is an interdisciplinary study that involves both analytical chemistry and an understanding of environmental science. Environmental chemists must first understand the chemicals and chemical reactions present in natural processes in the soil water and air. Sampling and analysis can then determine if human activities have contaminated the environment or caused harmful reactions to affect it.

Water quality is an important area of environmental chemistry. "Pure" water does not exist in nature; it always has some minerals or other substance dissolved in it. Water quality chemists test rivers, lakes and ocean water for characteristics such as dissolved oxygen, salinity, turbidity, suspended sediments, and pH. Water destined for human consumption must be free of harmful contaminants and may be treated with additives like fluoride and chlorine to increase its safety.

Agricultural Chemistry

Agricultural chemistry is concerned with the substances and chemical reactions that are involved with the production, protection and use of crops and livestock. It is a highly interdisciplinary field that relies on ties to many other sciences. Agricultural chemists may work with the Department of Agriculture, the Environmental Protection Agency, the Food and Drug Administration or for private industry. Agricultural chemists develop fertilizers, insecticides and herbicides necessary for large-scale crop production. They must also monitor how these products are used and their impacts on the environment. Nutritional supplements are developed to increase the productivity of meat and dairy herds.

Agricultural biotechnology is a fast-growing focus for many agricultural chemists. Genetically manipulating crops to be resistant to the herbicides used to control weeds in the fields requires detailed understanding of both the plants and the chemicals at the molecular level. Biochemists must understand genetics, chemistry and business needs to develop crops that are easier to transport or that have a longer shelf life.

Chemical Engineering

Chemical engineers research and develop new materials or processes that involve chemical reactions. Chemical engineering combines a background in chemistry with engineering and economics concepts to solve technological problems. Chemical engineering jobs fall into two main groups: industrial applications and development of new products.

Industries require chemical engineers to devise new ways to make the manufacturing of their products easier and more cost effective. Chemical engineers are involved in designing and operating processing plants, develop safety procedures for handling dangerous materials, and supervise the manufacture of nearly every product we use. Chemical engineers work to develop new products and processes in every field from pharmaceuticals to fuels and computer components.

Geochemistry

Geochemists combine chemistry and geology to study the makeup and interaction between substances found in the Earth. Geochemists may spend more time in field studies than other types of chemists. Many work for the U.S. Geological Survey or the Environmental Protection Agency in determining how mining operations and waste can affect water quality and the environment. They may travel to remote abandoned mines to collect samples and perform rough field evaluations, and then follow a stream through its watershed to evaluate how contaminants are moving through the system. Petroleum geochemists are employed by oil and gas companies to help find new energy reserves. They may also work on pipelines and oil rigs to prevent chemical reactions that could cause explosions or spills.

Forensic Chemistry

Forensic chemists capture and analyze the physical evidence left behind at a crime scene to help determine the identities of the people involved as well as to answer other vital questions regarding how and why the crime was carried out. Forensic chemists use a wide variety of analyzation methods, such as chromatography, spectrometry and spectroscopy.

Atoms and Molecules

Atom

The atom is the smallest unit of matter that is composed of three sub-atomic particles: the proton, the neutron, and the electron. Protons and neutrons make up the nucleus of the atom, a dense and positively charged core, whereas the negatively charged electrons can be found around the nucleus in an electron cloud.

Atomic Particles

Atoms consist of three basic particles: protons, electrons, and neutrons. The nucleus (center) of the atom contains the protons (positively charged) and the neutrons (no charge). The outermost regions of the atom are called electron shells and contain the electrons (negatively charged). Atoms have different properties based on the arrangement and number of their basic particles.

The hydrogen atom (H) contains only one proton, one electron, and no neutrons. This can be determined using the atomic number and the mass number of the element.

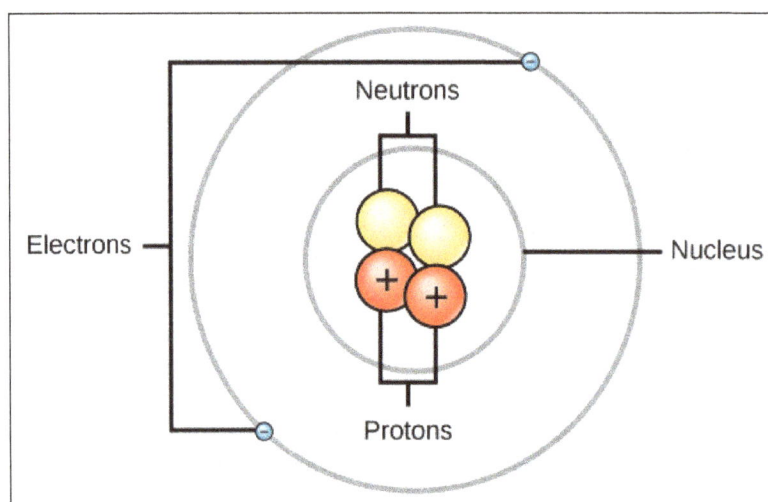

Structure of an atom: Elements, such as helium, depicted here, are made up of atoms. Atoms are made up of protons and neutrons located within the nucleus, with electrons in orbitals surrounding the nucleus.

Atomic Mass

Protons and neutrons have approximately the same mass, about 1.67×10^{-24} grams. Scientists define this amount of mass as one atomic mass unit (amu) or one Dalton. Although similar in mass, protons are positively charged, while neutrons have no charge. Therefore, the number of neutrons in an atom contributes significantly to its mass, but not to its charge.

Electrons are much smaller in mass than protons, weighing only 9.11×10^{-28} grams, or about 1/1800 of an atomic mass unit. Therefore, they do not contribute much to an element's overall atomic mass. When considering atomic mass, it is customary to ignore the mass of any electrons and calculate the atom's mass based on the number of protons and neutrons alone.

Electrons contribute greatly to the atom's charge, as each electron has a negative charge equal to the positive charge of a proton. Scientists define these charges as "+1" and "-1". In an uncharged, neutral atom, the number of electrons orbiting the nucleus is equal to the number of protons inside the nucleus. In these atoms, the positive and negative charges cancel each other out, leading to an atom with no net charge.

Protons, Neutrons, and Electrons			
	Charge	Mass (amu)	Location
Proton	+1	1	nucleus
Neutron	0	1	nucleus
Electron	−1	0	orbitals

Table shows protons, neutrons, and electrons. Both protons and neutrons have a mass of 1 amu and are found in the nucleus. However, protons have a charge of +1, and neutrons are uncharged. Electrons have a mass of approximately 0 amu, orbit the nucleus, and have a charge of -1.

Exploring Electron Properties: Compare the behavior of electrons to that of other charged particles to discover properties of electrons such as charge and mass.

Volume of Atoms

Accounting for the sizes of protons, neutrons, and electrons, most of the volume of an atom—greater than 99 percent—is, in fact, empty space. Despite all this empty space, solid objects do not just pass through one another. The electrons that surround all atoms are negatively charged and cause atoms to repel one another, preventing atoms from occupying the same space. These intermolecular forces prevent you from falling through an object like your chair.

Interactive: Build an Atom - Build an atom out of protons, neutrons, and electrons, and see how the element, charge, and mass change.

Atomic Number and Mass Number

The atomic number is the number of protons in an element, while the mass number is the number of protons plus the number of neutrons.

Atomic Number

Neutral atoms of an element contain an equal number of protons and electrons. The number of protons determines an element's atomic number (Z) and distinguishes one element from another. For example, carbon's atomic number (Z) is 6 because it has 6 protons. The number of neutrons can vary to produce isotopes, which are atoms of the same element that have different numbers of neutrons. The number of electrons can also be different in atoms of the same element, thus producing ions (charged atoms). For instance, iron, Fe, can exist in its neutral state, or in the +2 and +3 ionic states.

Mass Number

An element's mass number (A) is the sum of the number of protons and the number of neutrons. The small contribution of mass from electrons is disregarded in calculating the mass number. This approximation of mass can be used to easily calculate how many neutrons an element has by simply subtracting the number of protons from the mass number. Protons and neutrons both weigh about one atomic mass unit or amu. Isotopes of the same element will have the same atomic number but different mass numbers.

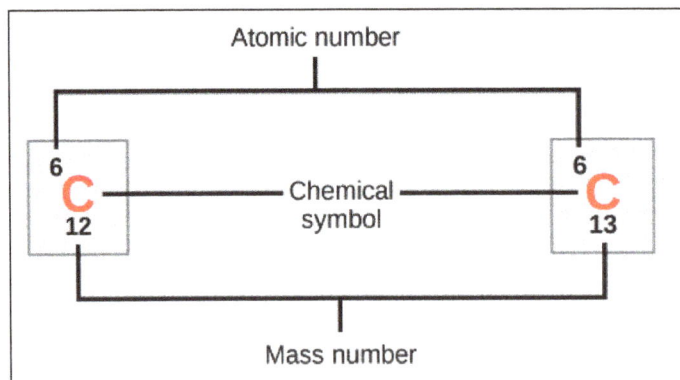

Atomic number, chemical symbol, and mass number: Carbon has an atomic number of six, and two stable isotopes with mass numbers of twelve and thirteen, respectively. Its average atomic mass is 12.11.

Scientists determine the atomic mass by calculating the mean of the mass numbers for its naturally-occurring isotopes. Often, the resulting number contains a decimal. For example, the atomic mass of chlorine (Cl) is 35.45 amu because chlorine is composed of several isotopes, some (the majority) with an atomic mass of 35 amu (17 protons and 18 neutrons) and some with an atomic mass of 37 amu (17 protons and 20 neutrons).

Given an atomic number (Z) and mass number (A), you can find the number of protons, neutrons, and electrons in a neutral atom. For example, a lithium atom (Z=3, A=7 amu) contains three protons (found from Z), three electrons (as the number of protons is equal to the number of electrons in an atom), and four neutrons ($7 - 3 = 4$).

Isotopes

Isotopes are various forms of an element that have the same number of protons but a different number of neutrons. Some elements, such as carbon, potassium, and uranium, have multiple naturally-occurring isotopes. Isotopes are defined first by their element and then by the sum of the protons and neutrons present:

- Carbon-12 (or ^{12}C) contains six protons, six neutrons, and six electrons; therefore, it has a mass number of 12 amu (six protons and six neutrons).

- Carbon-14 (or ^{14}C) contains six protons, eight neutrons, and six electrons; its atomic mass is 14 amu (six protons and eight neutrons).

While the mass of individual isotopes is different, their physical and chemical properties remain mostly unchanged.

Isotopes do differ in their stability. Carbon-12 (^{12}C) is the most abundant of the carbon isotopes, accounting for 98.89% of carbon on Earth. Carbon-14 (^{14}C) is unstable and only occurs in trace amounts. Unstable isotopes most commonly emit alpha particles (He^{2+}) and electrons. Neutrons, protons, and positrons can also be emitted and electrons can be captured to attain a more stable atomic configuration (lower level of potential energy) through a process called radioactive decay. The new atoms created may be in a high energy state and emit gamma rays which lowers the energy but alone does not change the atom into another isotope. These atoms are called radioactive isotopes or radioisotopes.

Radiocarbon Dating

Carbon is normally present in the atmosphere in the form of gaseous compounds like carbon dioxide and methane. Carbon-14 (^{14}C) is a naturally-occurring radioisotope that is created from atmospheric ^{14}N (nitrogen) by the addition of a neutron and the loss of a proton, which is caused by cosmic rays. This is a continuous process so more ^{14}C is always being created in the atmosphere. Once produced, the ^{14}C often combines with the oxygen in the atmosphere to form carbon dioxide. Carbon dioxide produced in this way diffuses in the atmosphere, is dissolved in the ocean, and is incorporated by plants via photosynthesis. Animals eat the plants and, ultimately, the radiocarbon is distributed throughout the biosphere.

In living organisms, the relative amount of ^{14}C in their body is approximately equal to the concentration of ^{14}C in the atmosphere. When an organism dies, it is no longer ingesting ^{14}C, so the ratio between ^{14}C and ^{12}C will decline as ^{14}C gradually decays back to ^{14}N. This slow process, which is called beta decay, releases energy through the emission of electrons from the nucleus or positrons.

After approximately 5,730 years, half of the starting concentration of ^{14}C will have been converted back to ^{14}N. This is referred to as its half-life, or the time it takes for half of the original concentration of an isotope to decay back to its more stable form. Because the half-life of ^{14}C is long, it is used to date formerly-living objects such as old bones or wood. Comparing the ratio of the ^{14}C concentration found in an object to the amount of ^{14}C in the atmosphere, the amount of the isotope that has not yet decayed can be determined. On the basis of this amount, the age of the material can be accurately calculated, as long as the material is believed to be less than 50,000 years old. This technique is called radiocarbon dating, or carbon dating for short.

Application of carbon dating: The age of carbon-containing remains less than 50,000 years old, such as this pygmy mammoth, can be determined using carbon dating.

Other elements have isotopes with different half-lives. For example, ^{40}K (potassium-40) has a half-life of 1.25 billion years, and ^{235}U (uranium-235) has a half-life of about 700 million years. Scientists often use these other radioactive elements to date objects that are older than 50,000 years (the limit of carbon dating). Through the use of radiometric dating, scientists can study the age of fossils or other remains of extinct organisms.

Molecule

A molecule is the smallest particle in a chemical element or compound that has the chemical properties of that element or compound. Molecules are made up of atoms that are held together by chemical bonds. These bonds form as a result of the sharing or exchange of electrons among atoms. The atoms of certain elements readily bond with other atoms to form molecules. Examples of such elements are oxygen and chlorine. The atoms of some elements do not easily bond with other atoms. Examples are neon and argon.

Molecules can vary greatly in size and complexity. The element helium is a one-atom molecule. Some molecules consist of two atoms of the same element. For example, O_2 is the oxygen molecule most commonly found in the earth's atmosphere; it has two atoms of oxygen. However, under certain circumstances, oxygen atoms bond into triplets (O_3), forming a molecule known as ozone. Other familiar molecules include water, consisting of two hydrogen atoms and one oxygen atom (H_2O), carbon dioxide, consisting of one carbon atom bonded to two oxygen atoms (CO_2), and sulfuric acid, consisting of two hydrogen atoms, one sulfur atom, and four oxygen atoms (H_2SO_4).

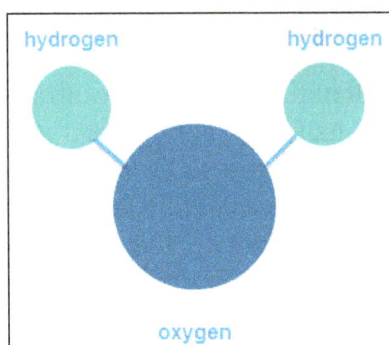

Water molecule.

Some molecules, notably certain proteins, contain hundreds or even thousands of atoms that join together in chains that can attain considerable lengths. Liquids containing such molecules sometimes behave strangely. For example, a liquid may continue to flow out of a flask from which some of it has been poured, even after the flask is returned to an upright position.

Molecules are always in motion. In solids and liquids, they are packed tightly together. In a solid, the motion of the molecules can be likened to rapid vibration. In a liquid, the molecules can move freely among each other, in a sort of slithering fashion. In a gas, the density of molecules is generally less than in a liquid or solid of the same chemical compound, and they move even more freely than in a liquid. For a specific compound in a given state (solid, liquid, or gas), the speed of molecular motion increases as the absolute temperature increases.

Laws of Chemical Combinations

There was tremendous progress in Chemical Sciences after 18th century. It arose out of an interest in the nature of heat and the way things burn. Major progress was made through the careful use of chemical balance to determine the change in mass that occurs in chemical reactions. The great French Chemist Antoine Lavoisier used the balance to study chemical reactions. He heated mercury in a sealed flask that contained air. After several days, a red substance mercury (II) oxide was produced. The gas remaining in the flask was reduced in mass. The remaining gas was neither able to support combustion nor life. The remaining gas in the flask was identified as nitrogen. The gas which combined with mercury was oxygen. Further he carefully performed the experiment by taking a weighed quantity of mercury (II) oxide. After strong heating, he found that mercury (II) oxide, red in colour, was decomposed into mercury and oxygen. He weighed both mercury and oxygen and found that their combined mass was equal to that of the mercury (II) oxide taken. Lavoisier finally came to the conclusion that in every chemical reaction, total masses of all the reactants is equal to the masses of all the products. This law is known as the law of conservation of mass.

There was rapid progress in science after chemists began accurate determination of masses of reactants and products. French chemist Claude Berthollet and Joseph Proust worked on the ratio (by mass) of two elements which combine to form a compound. Through a careful work, Proust demonstrated the fundamental law of definite or constant proportions in 1808. In a given chemical compound, the proportions by mass of the elements that compose it are fixed, independent of the origin of the compound or its mode of preparation.

In pure water, for instance, the ratio of mass of hydrogen to the mass of oxygen is always 1:8 irrespective of the source of water. In other words, pure water contains 11.11% of hydrogen and 88.89% of oxygen by mass whether water is obtained from well, river or from a pond. Thus, if 9.0 g of water are decomposed, 1.0 g of hydrogen and 8.0 g of oxygen are always obtained. Furthermore, if 3.0 g of hydrogen are mixed with 8.0 g of oxygen and the mixture is ignited, 9.0 g of water are formed and 2.0 g of hydrogen remains unreacted. Similarly sodium chloride contains 60.66% of chlorine and 39.34% of sodium by mass whether we obtained it from salt mines or by crytallising it from water of ocean or inland salt seas or synthesizing it from its elements sodium and chlorine. Of course, the key word in this sentence is 'pure'. Reproducible experimental results are highlights of scientific thoughts. In fact modern science is based on experimental findings. Reproducible results indirectly hint for a truth which is hidden. Scientists always worked for findings this truth and in this manner many theories and laws were discovered. This search for truth plays an important role in the development of science.

Dalton's Atomic Theory

The English scientist John Dalton was by no means the first person to propose the existence of atoms, as we have seen in the previous section, such ideas date back to classical times. Dalton's major contribution was to arrange those ideas in proper order and give evidence for the existence of atoms. He showed that the mass relationship expressed by Lavoisier and Proust (in the form of law of conservation of mass and law of constant proportions) could be interpreted most suitably by postulating the existence of atoms of the various elements.

In 1803, Dalton published a new system of chemical philosophy in which the following statements comprise the atomic theory of matter:

- Matter consists of indivisible atoms.

- All the atoms of a given chemical element are identical in mass and in all other properties.

- Different chemical elements have different kinds of atoms and in particular such atoms have different masses.

- Atoms are indestructible and retain their identity in chemical reactions.

- The formation of a compound from its elements occurs through the combination of atoms of unlike elements in small whole number ratio.

Dalton's fourth postulate is clearly related to the law of conservation of mass. Every atom of an element has a definite mass. Also in a chemical reaction there is rearrangement of atoms. Therefore after the reaction, mass of the product should remain the same. The fifth postulate is an attempt to explain the law of definite proportions. A compound is a type of matter containing the atoms of two or more elements in small whole number ratio. Because the atoms have definite mass, the compound must have the elements in definite proportions by mass.

John Dalton.

The Dalton's atomic theory not only explained the laws of conservations of mass and law of constant proportions but also predicted the new ones. He deduced the law of multiple proportions on the basis of his theory. The law states that when two elements form more than one compound, the masses of one element in these compound for a fixed mass of the other element are in the ratio of small whole numbers. For example, carbon and oxygen form two compounds: Carbon monoxide and carbon dioxide. Carbon monoxide contains 1.3321 g of oxygen for each 1.000g of carbon, whereas carbon dioxide contains 2.6642 g of oxygen for 1.0000 g of carbon. In other words, carbon dioxide contains twice the mass of oxygen as is contained in carbon monoxide (2.6642 g = 2 × 1.3321 g) for a given mass of carbon. Atomic theory explains this by saying that carbon dioxide contains twice as many oxygen atoms for a given number of carbon atoms as does carbon monoxide. The deduction of law of multiple proportions from atomic theory was important in convincing chemists of the validity of the theory.

Atomic Mass

Dalton gave the concept of atomic mass. According to him, atoms of the same element have same atomic masses but atoms of different elements have different atomic masses. Since Dalton could not weigh individual atoms, he measured relative masses of the elements required to form a compound. From this, he deduced relative atomic masses. For example, we can determine by experiment that 1.0000 g of hydrogen gas reacts with 7.9367 g of oxygen gas to form water. If we know formula of water, we can easily determine the mass of an oxygen atom relative to that of hydrogen atom.

Dalton did not have a way of determining the proportions of atoms of each element forming water during those days. He assumed the simplest possibility that atoms of oxygen and hydrogen were equal in number. From this assumption, it would follow that oxygen atom would have a mass that was 7.9367 times that of hydrogen atom. This in fact was not correct. We now know that in water number of hydrogen atoms is twice the number of oxygen atoms (formula of water being H_2O). Therefore, relative mass of oxygen atom must be $2 \times 7.9367 = 15.873$ times that of hydrogen atom. After Dalton, relative atomic masses of several elements were determined by scientists based on hydrogen scale. Later on, hydrogen based scale was replaced by a scale based on oxygen as it (oxygen) was more reactive and formed a large number of compounds.

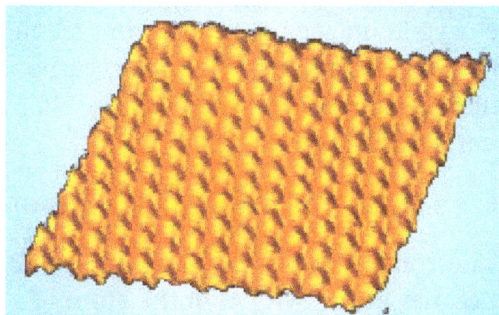

Image of Copper surface by STM technique Atom can be seen in magnified image of surface.

In 1961, C-12 (or $^{12}_{6}C$) atomic mass scale was adopted. This scale depends on measurement of atomic mass by an instrument called mass spectrometer. Mass spectrometer invented early in 20th century, allows us to determine atomic masses precisely. The masses of atoms are obtained by comparison with C-12 atomic mass scale. In fact C-12 isotope is chosen as standard and arbitrarily assigned a mass of exactly 12 atomic mass units. One atomic mass unit (amu), therefore, equals exactly one twelfth of mass of a carbon–12 atom, Atomic mass unit (amu) is nowa-days is written as unified mass unit and is denoted by the letter 'u'.

The relative atomic mass of an element expressed in atomic mass unit is called its atomic weight. Now-a-days we are using atomic mass in place of atomic weight.

Further, you have seen that Dalton proposed that masses of all atoms in an element are equal. But later on it was found that all atoms of naturally occurring elements are not of the same mass. We shall study about such atoms in the following section. Atomic masses that we generally use in our reaction or in chemical calculations are average atomic masses which depend upon relative abundance of isotopes of elements.

Isotopes and Atomic Mass

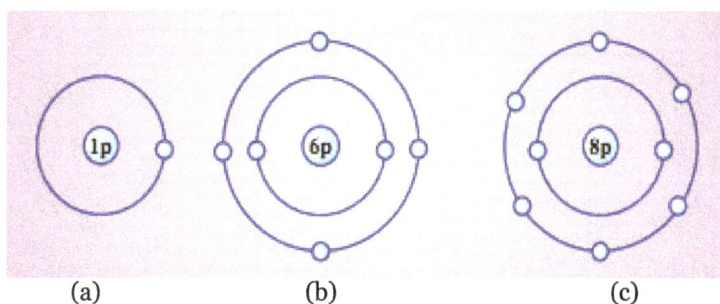

Arrangement of electrons around nucleus in (a) hydrogen, (b) carbon and (c) oxygen atoms.

Dalton considered an atom as an indivisible particle. Later researches proved that an atom consists of several fundamental particles such as: electrons, protons and neutrons. An electron is negatively charged and a proton is positively charged particle. Number of electrons and protons in an atom is equal. Since charge on an electron is equal and opposite to charge of a proton, therefore an atom is electrically neutral. Protons remain in the nucleus in the centre of the atom, and nucleus is surrounded by negatively charged electrons.

The number of protons in the nucleus is called atomic number denoted by Z. For example in figure, there are 8 protons in the oxygen nucleus, 6 protons in carbon nucleus and only one proton in hydrogen nucleus. Therefore atomic numbers of oxygen, carbon and hydrogen are 8,6 and 1 respectively. There are also neutral particles in the nucleus and they are called 'neutrons'. Mass of a proton and of a neutron is nearly the same.

Total mass of the nucleus = mass of protons + mass of neutrons

Total number of protons and neutrons is called mass number (A). By convention atomic number is written at the bottom of left corner of the symbol of the atom of a particular element and mass number is written at the top left corner. For example, symbol $^{12}_{6}C$ indicates that there is a total of 12 particles (nucleons) in the nucleus of a carbon atom, 6 of which are protons. Thus, there must be 12-6 = 6 neutrons. Similarly $^{16}_{8}O$ indicates 8 protons and 16 nucleons (8 protons + 8 neutrons). Since atom is electrically neutral, oxygen has 8 protons and 8 electrons in it. Further, atomic number (Z) differentiates the atom of one element from the atoms of the other elements.

An element may be defined as a substance where all the atoms have the same atomic number.

But the nuclei of all the atoms of a given element do not necessarily contain the same number of neutrons. For example, atoms of oxygen, found in nature, have the same number of protons which makes it different from other elements, but their neutrons (in nucleus) are different. This is the reason that the masses of atoms of the same element are different. For example, one type of oxygen atom contains 8 protons and 8 neutrons in one atom, second type 8 protons and 9 neutrons and the third type contains 8 protons and 10 neutrons. We represent these oxygen atoms as $^{16}_{8}O$, $^{17}_{8}O$ and $^{18}_{8}O$ respectively. Atoms of an element that have the same atomic number (Z) but different mass number (A) are called isotopes. In view of difference in atomic masses of the same element, we take average atomic masses of the elements. This is calculated on the basis of the abundance of the isotopes. Atomic masses of some elements are provided in table.

Table: Atomic mass of some common elements.

Elements	Symbol	Mass (u)	Elements	Symbol	Mass (u)
Aluminium	Al	26.93	Magnesium	Mg	24.31
Argon	Ar	39.95	Manganese	Mn	54.94
Arsenic	As	74.92	Mercury	Hg	200.59
Barium	Ba	137.34	Neon	Ne	20.18
Boron	B	10.81	Nickel	Ni	58.71
Bromine	Br	79.91	Nitrogen	N	14.01
Caesium	Cs	132.91	Oxygen	O	16.00
Calcium	Ca	40.08	Phosphorus	P	30.97
Carbon	C	12.01	Platinum	Pt	195.09

Chlorine	Cl	35.45	Potassium	K	39.1
Chromium	Cr	52.00	Radon	Rn	(222)
Cobalt	Co	58.93	Silicon	Si	23.09
Copper	Cu	63.56	Silver	Ag	107.87
Fluorine	F	19.00	Sodium	Na	23.00
Gold	Au	196.97	Sulphur	S	32.06
Helium	He	4.00	Tin	Sn	118.69
Hydrogen	H	1.008	Titanium	Ti	47.88
Iodine	I	126.90	Tungsten	W	183.85
Iron	Fe	55.85	Uranium	U	238.03
Lead	Pb	207.19	Vanadium	V	50.94
Lithium	Li	6.94	Xenon	Xe	131.30
			Zinc	Zn	65.37

Mole Concept

When we mix two substances, we get one or more new substances. For example, when we mix hydrogen and oxygen and ignite the mixture, we get a new substancewater. This can be represented in the form of a chemical equation,

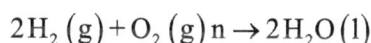

$$2H_2(g) + O_2(g)n \rightarrow 2H_2O(l)$$

In the above equation, 2 molecules (four atoms) of hydrogen react with 1 molecule (2 atoms) of oxygen and give two molecules of water. We always like to know how many atoms/molecules of a particular substance would react with atoms/molecules of another substance in a chemical reaction. No matter how small they are. The solution to this problem is to have a convenient unit. Would you not like to have a convenient unit? Definitely a unit for counting of atoms/molecules present in a substance will be desirable and convenient as well. This chemical counting unit of atoms and molecules is called mole.

The word mole was, apparently introduced in about 1896 by Wilhelm Ostwald who derived the term from the Latin word 'moles' meaning a 'heap' or 'pile'. The mole whose symbol is 'mol' is the SI (international system) base unit for measuring amount of substance. It is defined as follows:

> "A mole is the amount of substance that contains as many elementary entities (atoms, mol-ecules, formula unit or other fundamental particles) as there are atoms in exactly 0.012 kg of carbon-12 isotope."

In simple words, mole is the number of atoms in exactly 0.012 kg (12 grams) of C-12. Although mole is defined in terms of carbon atoms but the unit is applicable to any substance just as 1 dozen means 12 or one gross means 144 of anything. Mole is scientist's counting unit like dozen or gross. By using mole, scientists (particularly chemists) count atoms and molecules in a given substance. Now it is experimentally found that the number of atoms contained in exactly 12 g of C-12 is 602,200 000 000 000 000 000 000 or 6.022×10^{23}. This number is called Avogadro's number in honour of Amedeo Avogadro, an Italian lawyer and physicist. When this number is divided by 'mole' it becomes a constant and is known as Avogadro's constant denoted by symbol, NA = 6.02×10^{23} mol^{-1}. We have seen that,

Atomic mass of C = 12 u

Atomic mass of He = 4 u

We can see that one atom of carbon is three times as heavy as one atom of helium. On the same logic 100 atoms of carbon are three times as heavy as 100 atoms of helium. Similarly 6.02×10^{23} atoms of carbon are three times as heavy as 6.02×10^{23} atoms of helium. But 6.02×10^{23} atoms of carbon weigh 12 g, therefore 6.02×10^{23} atoms of helium will weigh 1/3 × 12g = 4 g. We can take a few more examples of elements and can calculate the mass of one mole atoms of that element.

Molar Mass

Mass of one mole of a substance is called its molar mass. A substance may be an element or a compound. Mass of one mole atoms of oxygen means mass of 6.02×10^{23} atoms of oxygen. It is found that one mole atoms of oxygen weighs 16.0 g. When we say one mole molecules of oxygen that means 6.02×10^{23} molecules of oxygen (O_2). One mole molecules of oxygen will weigh 32.0 g. Thus,

Mass of one mole atoms of oxygen = 16 g mol^{-1}

Mass of one mole molecules of oxygen = 32 g mol^{-1}

When it is not clear whether we are asking for one mole of atoms or one mole of molecules then we take natural form of that substance. For example, one mole of oxygen means one mole of oxygen molecules as oxygen occurs in the form of molecules in nature. In case of compounds, the same logic is applicable. For example, one mole of water means one mole molecules of water which weighs 18 g. Numerically one mole of a substance is equal to atomic or molecular mass of that substance expressed in grams.

Remember, molar mass is always expressed in the unit of g/mol or g mol^{-1}.

For example,

Molar mass of nitrogen (N2) = 28 g mol^{-1}

Molar mass of chlorine (Cl2) = 71 g mol^{-1}

Provides molecular and molar mass of a few common substances.

Molecular and Molar Masses

Formula	Molecular mass (u)	Molar mass(g/mol)
O_2(oxygen)	32.0	32.0
Cl_2(chlorine)	71.0	71.0
P_4 (phosphorus)	123.9	123.9
CH_4 (methane)	16.00	16.0
NH_3 (ammonia)	17.0	17.0

HCl (hydrochloric acid gas)	36.5	36.5
CO_2 (carbon dioxide)	44.0	44.0
SO_2 (sulphur dioxide)	64.0	64.0
C_2H_5OH (ethyl alcohol)	46.0	46.0
C_6H_6 (benzene)	78.0	78.00

Writing Chemical Formula of Compounds

As you are aware, a compound is made of two or more than two elements combined in a definite proportion by mass (law of constant proportions). Thus, the number of combining atoms in a compound is fixed. The elements are represented by their symbols (e.g. H for hydrogen, Na for sodium). Similarly a compound is also represented by a shorthand notation known as formula or chemical formula. The formula of a compound indicates (i) elements constituting the compound and (ii) the number of each constituent element. In other word, the formula of a compound also represents its chemical composition. The atoms of elements constituting a compound are indicated by their symbols and their number is indicated as a subscript on the right hand bottom of the symbol. For example, in the formula of water, H_2O, two atoms of hydrogen are indicated as subscript '2', while oxygen is shown without writing any subscript, which means that the number of oxygen atom is just one.

Valency and Formulation

Every element has a definite capacity to combine with other elements. This combining capacity of an element is called its valency. You will learn very soon that this combining capacity of elements depends on the electronic configuration of elements. Valencies of a few elements:

Valency of Elements

Elements	Symbol	Valency	Elements	Symbol	Valency
Hydrogen	H	1	Phosphorus	P	5
Oxygen	O	2	Sodium	Na	1
Carbon	C	4	Magnesium	Mg	2
Nitrogen	N	3	Calcium	Ca	2
Chlorine	Cl	1	Aluminium	Al	3
Bromine	Br	1	Iron	Fe	2
Iodine	I	1	Barium	Ba	2

Most of the simple compounds are made of two elements. Such compounds are called binary compounds. It is easy to write formula of such compounds. When a metal combines with a non-metal, the symbol of the metal-element is written on the left hand side and that of the non-metal element on right hand side. (If both are non-metal, we write more electronegative* element on the right hand side). In naming a compound, the first element is written as such and the name of the second element i.e. more electronegative element, changes its ending to 'ide'. For writing chemical formula, we have to write valencies as shown below and then cross over the valencies of the combining atoms. Formula of the compounds resulting from carbon and chlorine, hydrogen and oxygen, and hydrogen and chlorine can be written as follows:

Element	C Cl	H O	H Cl
Valency	4 1	1 2	1 1
Formula	CC_4	H_2O	HCl

Some other examples for writing formula of compounds CaO, NaCl and NH_3 can also be taken for more clarity.

Element	Ca O	Na Cl	N H
Valency	2 2	1 1	3 1
Formula	CaO	NaCl	NH_3

Thus, we can write formulas of various compounds if we know elements constituting them and their valencies.

Valency, as mentioned, depends on the electronic configuration and nature of the elements. Sometimes an element shows more than one type of valency. We say element has variable valency. For example nitrogen forms several oxides : N_2O, N_2O_2, N_2O_3, N_2O_4 and N_2O_5. If we take valency of oxygen equal to 2, then valency of nitrogen in the oxides will be 1,2,3,4 and 5 respectively. Valencies are not always fixed. Similar to nitrogen, phosphorus also shows valencies 3 and 5 as reflected in compounds PBr_3 and P_2O_5. In these compounds, there are more than one atom. In such cases, number of atoms is indicated by attaching a numerical prefix (mono, di, tri, etc.)

Table: Numerical Prefixes.

Number of atoms	Prefix	Example
1	Mono	carbon monoxide, CO
2	Di	carbon dioxide, CO_2
3	Tri	phosphorus trichloride, PCl_3
4	Tetra	carbon tetrachloride, CCl_4
5	Penta	Dinitrogen pentoxide, N_2O_5

Here you would notice that '–o' or '–a' at the end of the prefix is often dropped before another vowel, e.g. monoxide, pentoxide. There is no gap between numerical prefix and the name of the element. The prefix mono is usually dropped for the first element. When hydrogen is the first element in the formula, no prefix is added before hydrogen irrespective of the number. For example, the compound H_2S is named as hydrogen sulphide and not as dihydrogen sulphide.

Thus, we have seen that writing formula of a binary compound is relatively easy. However, when we have to write formula of a compound which involves more than two elements (i.e. of a polyatomic molecule), it is somewhat a cumbersome task. In the following section we shall consider formulation of more difficult compounds.

You will learn later on that there are basically two types of compounds: covalent compounds and ionic or electrovalent compounds. H_2O and NH_3 are covalent compounds. NaCl and MgO are ionic compounds. An ionic compound is made of two charged constituents. One positively charged and other negatively charged. In case of NaCl, there are two ions: Na^+ and Cl^- ion. Charge of these ions in case of electrovalent compound is used for writing formula. It is easy to write formula of an ionic compound only if there is one metal and one non-metal as in the case of NaCl and MgO. If there are more than two elements in an ionic compound, formulation will be a little difficult and in that situation we should know charge of cations and anions.

Formulation of Ionic Compounds

Formulation of an ionic compound is easy when we know charge of cation and anion. Remember, in an ionic compound, sum of the charge of cation and anion should be equal to zero. A few examples of cations and anions with their charges are provided.

Table: Charges of some common cations and anions which form ionic compounds.

Anions	Charge	Cations	Charge
Chloride ion, Cl^-	-1	Potassium ion, K^+	+1
Nitrate ion, NO_3^-	-1	Sodium ion, Na^+	+1
Hydroxide ion, OH^-	-1	Ammonium ion, NH_4^+	+1
Bicarbonate ion, HCO_3^-	-1	Magnesium ion, Mg_2^+	+2
Nitrite ion, NO_2^-	-1	Calcium ion, Ca_2^+	+2
Acetate ion, CH_3COO^-	-1	Lead ion, Pb_2^+	+2
Bromide ion, Br^-	-1	Iron ion (ous), Fe_2^+	+2
Iodide ion, I^-	-1	Zinc ion, Zn_2^+	+2
Sulphite ion, SO_3^{2-}	-2	Copper ion (cupric), Cu_2^+	+2
Carbonate ion, CO_3^{2-}	-2	Mercury ion (Mercuric), Hg_2^+	+2
Sulphide ion, S_2^-	-2	Iron (ic) ion, Fe_3^+	+3
Phosphate ion, PO_4^{3-}	-3	Aluminium ion, Al_3^+	+3
		Potassium ion, K^+	+1
		Sodium ion, Na^+	+1

Suppose you have to write formula of sodium sulphate which is made of Na^+ and SO_4^{2-} ions. For this the positive and negative charge can be crossed over to give subscripts. The purpose of this crossing over of charges is to find the number of ions required to equate the number of positive and negative charges.

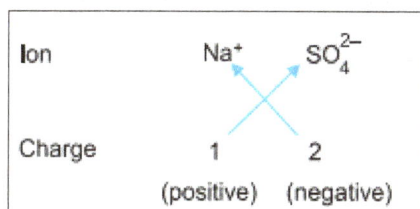

This gives the formula of sodium sulphate as Na_2SO_4. We can check the charge balance as follows:

$$\left.\begin{array}{rll} 2\,Na^+ = & 2\times(+1) = & +2 \\ 1SO_4^{2-} = & 1\times(-2) = & -2 \end{array}\right\} = 0$$

Thus the compound, Na_2SO_4 is electrically neutral.

Now it is clear that digit showing charge of cation goes to anion and digit showing charge of anion goes to cation. For writing formula of calcium phosphate we take charge of each ion into consideration and write the formula as discussed above.

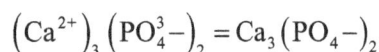

$$\left(Ca^{2+}\right)_3\left(PO_4^{3-}\right)_2 = Ca_3\left(PO_4-\right)_2$$

Writing formula of a compound comes only by practice, therefore write formulas of as many ionic compounds as possible based on the guidelines given above.

Measurements in Chemistry

Measurements provide the macroscopic information that is the basis of most of the hypotheses, theories, and laws that describe the behavior of matter and energy in both the macroscopic and microscopic domains of chemistry. Every measurement provides three kinds of information: the size or magnitude of the measurement (a number); a standard of comparison for the measurement (a unit); and an indication of the uncertainty of the measurement. While the number and unit are explicitly represented when a quantity is written, the uncertainty is an aspect of the measurement result that is more implicitly represented and will be discussed later.

The number in the measurement can be represented in different ways, including decimal form and scientific notation. (Scientific notation is also known as exponential notation;) For example, the maximum takeoff weight of a Boeing 777-200ER airliner is 298,000 kilograms, which can also be written as 2.98×10^5 kg. The mass of the average mosquito is about 0.0000025 kilograms, which can be written as 2.5×10^{-6} kg.

Units, such as liters, pounds, and centimeters, are standards of comparison for measurements. When we buy a 2-liter bottle of a soft drink, we expect that the volume of the drink was measured, so it is two times larger than the volume that everyone agrees to be 1 liter. The meat used to prepare a 0.25-pound hamburger is measured so it weighs one-fourth as much as 1 pound. Without units, a number can be meaningless, confusing, or possibly life threatening. Suppose a doctor prescribes phenobarbital to control a patient's seizures and states a dosage of "100" without specifying units. Not only will this be confusing to the medical professional giving the dose, but the consequences can be dire: 100 mg given three times per day can be effective as an anticonvulsant, but a single dose of 100 g is more than 10 times the lethal amount.

We usually report the results of scientific measurements in SI units, an updated version of the metric system, using the units listed in table. Other units can be derived from these base units. The standards for these units are fixed by international agreement, and they are called the International

System of Units or SI Units. SI units have been used by the United States National Institute of Standards and Technology (NIST) since 1964.

Table: Base Units of the SI System.

Property Measured	Name of Unit	Symbol of Unit
length	meter	m
mass	kilogram	kg
time	second	s
temperature	kelvin	K
electric current	ampere	A
amount of substance	mole	mol
luminous intensity	candela	cd

Sometimes we use units that are fractions or multiples of a base unit. Ice cream is sold in quarts (a familiar, non-SI base unit), pints (0.5 quart), or gallons (4 quarts). We also use fractions or multiples of units in the SI system, but these fractions or multiples are always powers of 10. Fractional or multiple SI units are named using a prefix and the name of the base unit. For example, a length of 1000 meters is also called a kilometer because the prefix kilo means "one thousand," which in scientific notation is 10^3 (1 kilometer = 1000 m = 10^3 m). The prefixes used and the powers to which 10 are raised are listed in table.

Table: Common Unit Prefixes.

Prefix	Symbol	Factor	Example
femto	f	10^{-15}	1 femtosecond (fs) = 1×10^{-15} m (0.000000000001 m)
pico	p	10^{-12}	1 picometer (pm) = 1×10^{-12} m (0.000000000001 m)
nano	n	10^{-9}	4 nanograms (ng) = 4×10^{-9} g (0.000000004 g)
micro	μ	10^{-6}	1 microliter (μL) = 1×10^{-6} L (0.000001 L)
milli	m	10^{-3}	2 millimoles (mmol) = 2×10^{-3} mol (0.002 mol)
centi	c	10^{-2}	7 centimeters (cm) = 7×10^{-2} m (0.07 m)
deci	d	10^{-1}	1 deciliter (dL) = 1×10^{-1} L (0.1 L)
kilo	k	10^3	1 kilometer (km) = 1×10^3 m (1000 m)
mega	M	10^6	3 megahertz (MHz) = 3×10^6 Hz (3,000,000 Hz)
giga	G	10^9	8 gigayears (Gyr) = 8×10^9 yr (8,000,000,000 Gyr)
tera	T	10^{12}	5 terawatts (TW) = 5×10^{12} W (5,000,000,000,000 W)

SI Base Units

The initial units of the metric system, which eventually evolved into the SI system, were established in France during the French Revolution. The original standards for the meter and the kilogram were adopted there in 1799 and eventually by other countries.

Length

The standard unit of length in both the SI and original metric systems is the meter (m). A meter was originally specified as 1/10,000,000 of the distance from the North Pole to the equator. It is now defined as the distance light in a vacuum travels in 1/299,792,458 of a second. A meter is about 3 inches longer than a yard; one meter is about 39.37 inches or 1.094 yards. Longer distances are often reported in kilometers (1 km = 1000 m = 10^3 m), whereas shorter distances can be reported in centimeters (1 cm = 0.01 m = 10^{-2} m) or millimeters (1 mm = 0.001 m = 10^{-3} m).

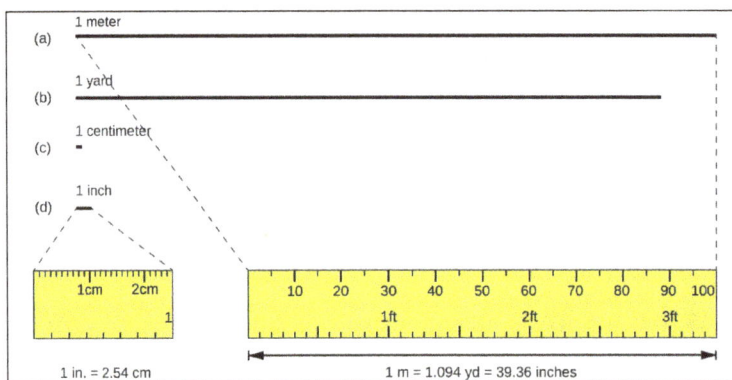

The relative lengths of 1 m, 1 yd, 1 cm, and 1 in. are shown (not actual size), as well as comparisons of 2.54 cm and 1 in., and of 1 m and 1.094 yd.

Mass

The standard unit of mass in the SI system is the kilogram (kg). A kilogram was originally defined as the mass of a liter of water (a cube of water with an edge length of exactly 0.1 meter). It is now defined by a certain cylinder of platinum-iridium alloy, which is kept in France. Any object with the same mass as this cylinder is said to have a mass of 1 kilogram. One kilogram is about 2.2 pounds. The gram (g) is exactly equal to 1/1000 of the mass of the kilogram (10^{-3} kg).

This replica prototype kilogram is housed at the National Institute of Standards and Technology (NIST) in Maryland.

Temperature

Temperature is an intensive property. The SI unit of temperature is the kelvin (K). The IUPAC convention is to use kelvin (all lowercase) for the word, K (uppercase) for the unit symbol, and neither the word "degree" nor the degree symbol (°). The degree Celsius (°C) is also allowed in the SI system, with both the word "degree" and the degree symbol used for Celsius measurements. Celsius degrees are the same magnitude as those of kelvin, but the two scales place their zeros in different places. Water freezes at 273.15 K (0 °C) and boils at 373.15 K (100 °C) by definition, and normal human body temperature is approximately 310 K (37 °C).

Time

The SI base unit of time is the second (s). Small and large time intervals can be expressed with the appropriate prefixes; for example, 3 microseconds = 0.000003 s = 3×10^{-6} and 5 megaseconds = 5,000,000 s = 5×10^{6} s. Alternatively, hours, days, and years can be used.

Derived SI Units

We can derive many units from the seven SI base units. For example, we can use the base unit of length to define a unit of volume, and the base units of mass and length to define a unit of density.

Volume

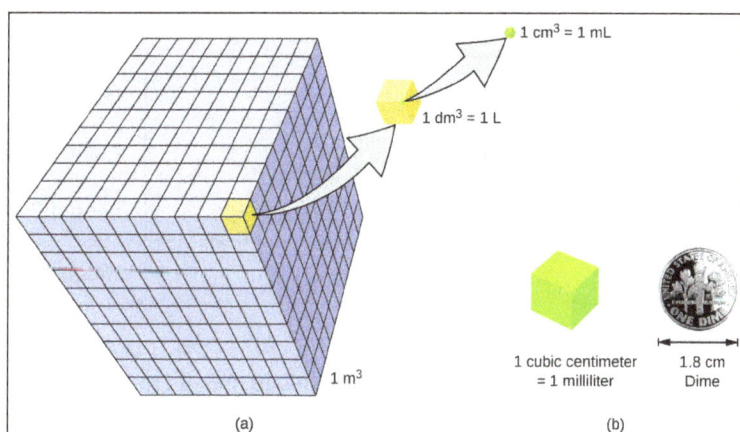

(a) The relative volumes are shown for cubes of 1 m³, 1 dm³ (1 L), and 1 cm³ (1 mL) (not to scale).
(b) The diameter of a dime is compared relative to the edge length of a 1-cm³ (1-mL) cube.

Volume is the measure of the amount of space occupied by an object. The standard SI unit of volume is defined by the base unit of length. The standard volume is a cubic meter (m³), a cube with an edge length of exactly one meter. To dispense a cubic meter of water, we could build a cubic box with edge lengths of exactly one meter. This box would hold a cubic meter of water or any other substance.

A more commonly used unit of volume is derived from the decimeter (0.1 m, or 10 cm). A cube with edge lengths of exactly one decimeter contains a volume of one cubic decimeter (dm³). A liter (L) is the more common name for the cubic decimeter. One liter is about 1.06 quarts.

A cubic centimeter (cm³) is the volume of a cube with an edge length of exactly one centimeter. The abbreviation cc (for cubic centimeter) is often used by health professionals. A cubic centimeter is also called a milliliter (mL) and is 1/1000 of a liter.

Density

We use the mass and volume of a substance to determine its density. Thus, the units of density are defined by the base units of mass and length.

The density of a substance is the ratio of the mass of a sample of the substance to its volume. The SI unit for density is the kilogram per cubic meter (kg/m^3). For many situations, however, this as an inconvenient unit, and we often use grams per cubic centimeter (g/cm^3) for the densities of solids and liquids, and grams per liter (g/L) for gases. Although there are exceptions, most liquids and solids have densities that range from about 0.7 g/cm^3 (the density of gasoline) to 19 g/cm^3 (the density of gold). The density of air is about 1.2 g/L. table shows the densities of some common substances.

Table: Densities of Common Substances.

Solids	Liquids	Gases (at 25 °C and 1 atm)
ice (at 0 °C) 0.92 g/cm^3	water 1.0 g/cm^3	dry air 1.20 g/L
oak (wood) 0.60–0.90 g/cm^3	ethanol 0.79 g/cm^3	oxygen 1.31 g/L
iron 7.9 g/cm^3	acetone 0.79 g/cm^3	nitrogen 1.14 g/L
copper 9.0 g/cm^3	glycerin 1.26 g/cm^3	carbon dioxide 1.80 g/L
lead 11.3 g/cm^3	olive oil 0.92 g/cm^3	helium 0.16 g/L
silver 10.5 g/cm^3	gasoline 0.70–0.77 g/cm^3	neon 0.83 g/L
gold 19.3 g/cm^3	mercury 13.6 g/cm^3	radon 9.1 g/L

While there are many ways to determine the density of an object, perhaps the most straightforward method involves separately finding the mass and volume of the object, and then dividing the mass of the sample by its volume. In the following example, the mass is found directly by weighing, but the volume is found indirectly through length measurements.

$$\text{density} = \frac{\text{volume}}{\text{mass}}.$$

Chemical Bonding

When two atoms of same or different elements approach each other, the energy of the combination of the atoms becomes less than the sum of the energies of the two separate atoms at a large distance. We say that the two atoms have combined or a bond is formed between the two. The bond is called a chemical bond. Thus a chemical bond may be visualised as an effect that leads to the decrease in the energy. The combination of atoms leads to the formation of a molecule that has distinct properties different from that of the constituent atoms.

A question arises, " How do atoms achieve the decrease in energy to form the bond". The answer lies in the electronic configuration. As you are aware, the noble gases do not react with other elements to form compounds. This is due to their stable electronic configuration with eight electrons (two in case of helium) in their outermost shells. The formation of a bond between two atoms may be visualised in terms of their acquiring stable electronic configurations. That is when two atoms (other than that of noble gases) combine they will do so in such a way that they attain an electronic configuration of the nearest noble gas.

The stable electronic configuration of the noble gases can be achieved in a number of ways; by losing, gaining or sharing of electrons. Accordingly, there are different types of chemical bonds, like,

- Ionic or electrovalent bond

- Covalent bond

- Co-ordinate covalent bond

In addition to these we have a special kind of bond called hydrogen bond. Let us discuss about different types of bonds, their formation and the properties of the compounds so formed.

Ionic or Electovalent Bond

According to Kossel's theory, in the process of formation of ionic bond the atoms acquire the noble gas electronic configuration by the gain or loss of electrons. Let us consider the formation of NaCl in terms of Kossel's Theory.

The electronic configuration of sodium atom (atomic number 11) is 2,8,1. Since it is highly electropositive, it readily loses an electron to attain the stable configuration of the nearest noble gas (neon) atom. It becomes a positively charged sodium cation (Na^+) in the process,

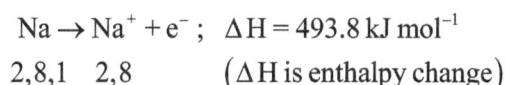

$$Na \rightarrow Na^+ + e^- \; ; \quad \Delta H = 493.8 \, kJ \, mol^{-1}$$
$$2,8,1 \quad 2,8 \qquad \left(\Delta H \text{ is enthalpy change} \right)$$

On the other hand, a chlorine atom (electronic configuration: 2,8,7) requires one electron to acquire the stable electronic arrangement of an argon atom. It becomes a negatively charged chloride anion (Cl^-) in the process.

$$Cl + \quad e^- \rightarrow Cl^- \; ; \qquad \Delta H = -379.5 \, kJ \, mol^{-1}$$
$$2,8,7 \quad 2,8,8$$

According to Kossel's theory, there is a transfer of one electron from sodium atom to chlorine atom and both the atoms attain noble gas configuration.

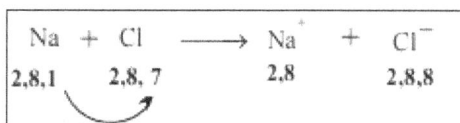

$$\boxed{\begin{array}{ccccccc} Na & + & Cl & \longrightarrow & Na^+ & + & Cl^- \\ 2,8,1 & & 2,8,7 & & 2,8 & & 2,8,8 \end{array}}$$

The positively charged sodium ion and the negatively charged chloride ion are held together by electrostatic attractions. The bond so formed is called an electrovalent or an ionic bond. Thus the ionic bond can be visualised as the electrostatic force of attraction that holds the cation and anion together. The compounds so formed are termed as ionic or electrovalent compounds.

Energetics of Ionic Compound Formation

We have just described the formation of an ionic compound (NaCl) as a result of transfer of electrons as proposed by Kossel. You may raise a question here that when more energy is required (ionisation energy) to form a sodium ion from sodium atom,than that released (electron affinity) in the formation of chloride ion from chlorine atom then how do we say that the formation of NaCl is accompanied by a decrease in energy? Your question is quite justified but let us assure you that there is no anomaly. Let us look at the whole process somewhat closely to clarify your doubts.

The formation of NaCl from sodium and chlorine can be broken down into a number of steps as:

- Sublimation of solid sodium to gaseous sodium atoms.

   $Na(s) \longrightarrow Na(g) ;$ $\Delta H = 108.7 \text{ kJ mol}^{-1}$

- Ionization of gaseous sodium atom to give sodium ion.

   $Na(g) \longrightarrow Na^+(g) + e^- ;$ $\Delta H = 493.8 \text{ kJ mol}^{-1}$

- Dissociation of gaseous chlorine molecule into chlorine atoms.

   $\frac{1}{2}Cl_2(g) \rightarrow Cl(g);$ $\Delta H = -120.9 \text{ kJ mol}^{-1}$

- Conversion of gaseous chlorine atom to chloride ion (addition of electron).

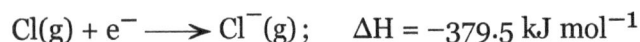

   $Cl(g) + e^- \longrightarrow Cl^-(g);$ $\Delta H = -379.5 \text{ kJ mol}^{-1}$

- Formation of NaCl from sodium and chloride ions.(Crystal or lattice formation).

   $Na^+(g) + Cl(g) \longrightarrow Na^+ Cl(s);$ $\Delta H = -754.8 \text{ kJ mol}^{-1}$

The energy released in this step is lattice energy.

The net reaction would be,

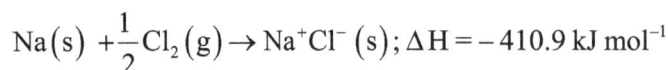

$$Na(s) + \frac{1}{2}Cl_2(g) \rightarrow Na^+Cl^-(s); \Delta H = -410.9 \text{ kJ mol}^{-1}$$

The overall energy change can be computed by taking the sum of all the energy changes:

$$\Delta H = (180.7 + 493.8 + 120.9 - 379.5 - 754.8) = -410.9 \text{ kJ mol}^{-1}$$

Thus we see that the net process of formation of NaCl from sodium and chlorine is accompanied by a large decrease in the energy. The approach we have just followed is based on the law of conservation of energy and is known as Born-Haber cycle.

Of the five different types of energies involved, two (sublimation and dissociation energies) are generally have low values than the rest. Therefore, the three energy terms i.e., ionization energy, electron affinity and lattice energy are important in determining the formation of an ionic compound. On the basis of the above discussion we can say that the formation of an ionic compound is favoured by:

- Low ionisation energy of the metal.

- High electron affinity of the other element (non-metal).

- High lattice energy.

Characteristic Properties of Ionic Compounds

- These exist as crystalline solids in which the ions are arranged in a regular three dimensional structure. The ionic compounds are generally hard and brittle in nature.

- These compounds have high melting and boiling points due to strong electrostatic interactions between the ions.

- These are generally soluble in water and less soluble in non-polar solvents like ether, alcohol, etc.

- These conduct electricity when in molten state or in aqueous solutions.

Kossel's theory explains bonding quite well but only for a small class of solids composed of electropositive elements of Group 1 and 2 with highly electronegative elements. Secondly, this theory is incapable of explaining the formation of compounds like, SO_2 or O_2 etc. For example in case of O_2, there is no reason to expect that one atom of oxygen would lose two electrons while the other accepts them. The problem was solved by Lewis theory of covalent bonding.

Covalent Bond

Like Kossel, Lewis also assumed that atoms attain noble gas electronic configuration in the process of bond formation. However, the way the noble gas electronic configuration is achieved, is different. Lewis proposed that this is achieved by "sharing of a pair of electrons" between the two atoms. Both the atoms contribute an electron each to this pair. For example, two hydrogen atoms form a molecule by sharing a pair of electrons. If electrons are indicated as dots, formation of hydrogen molecule can be shown as:

$$H. + .H \rightarrow H:H \rightarrow H-H$$

This shared pair of electrons contributes towards the stability of both the atoms and is said to be responsible for 'bonding' between the two atoms. Such a bond is called covalent bond and the compounds so obtained are called covalent compounds. In the process of suggesting the process of chemical bonding Lewis provided a very convenient way of representing bonding in simple molecules. This is called Lewis electron-dot structures or simply Lewis structures.

In Lewis structure each element is represented by a Lewis symbol. This symbol consists of the normal chemical symbol of the element surrounded by number of dots representing the electrons in the valence shell. Since the electrons are represented by dots, these are called electron-dot structures. The Lewis symbols of some elements are as:

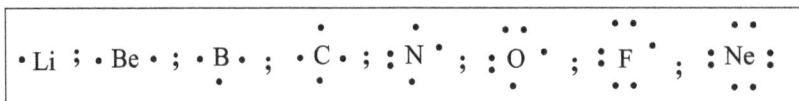

$$\cdot Li \; ; \; \cdot Be \cdot \; ; \; \cdot \overset{\cdot}{\underset{\cdot}{B}} \cdot \; ; \; \cdot \overset{\cdot}{\underset{\cdot}{C}} \cdot \; ; \; :\overset{\cdot}{N} \cdot \; ; \; :\overset{\cdot\cdot}{\underset{\cdot}{O}} \cdot \; ; \; :\overset{\cdot\cdot}{\underset{\cdot\cdot}{F}} \cdot \; ; \; :\overset{\cdot\cdot}{\underset{\cdot\cdot}{Ne}}:$$

You may note here that while writing the Lewis symbols, single dots are placed first on each side of the chemical symbol then they are paired up. The Lewis structure of a molecule is written in terms of these symbols.

In terms of Lewis symbols the ionic bond formation in NaCl can be represented as:

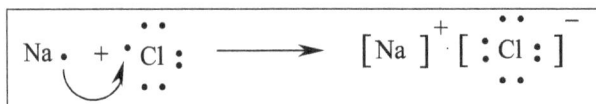

$$Na \cdot \; + \; \overset{\cdot\cdot}{\underset{\cdot\cdot}{Cl}}: \longrightarrow [Na]^+ \; [:\overset{\cdot\cdot}{\underset{\cdot\cdot}{Cl}}:]^-$$

and the covalent bond formation in HCl is represented as:

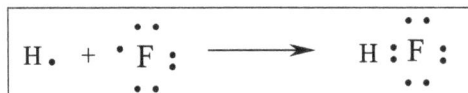

$$H \cdot \; + \; \cdot \overset{\cdot\cdot}{\underset{\cdot\cdot}{F}}: \longrightarrow H : \overset{\cdot\cdot}{\underset{\cdot\cdot}{F}}:$$

Sometimes the electrons contributed by different atoms are represented by different symbols. For example, formation of HF may also be shown as:

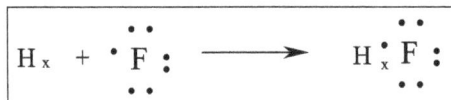

$$H_x \; + \; \cdot \overset{\cdot\cdot}{\underset{\cdot\cdot}{F}}: \longrightarrow H \overset{\cdot}{\underset{x}{}} \overset{\cdot\cdot}{\underset{\cdot\cdot}{F}}:$$

In this case the hydrogen electron is shown as a cross while the electrons of fluorine are represented by dots. There is no difference between electrons; it is just a presentation for the sake of convenience.

In terms of Lewis structures the formation of a chlorine molecule from two chlorine atoms may be represented as:

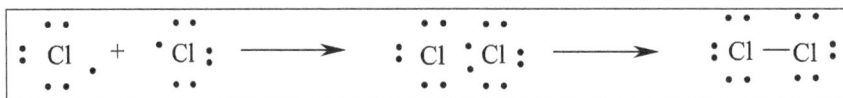

$$:\overset{\cdot\cdot}{\underset{\cdot\cdot}{Cl}} \cdot \; + \; \cdot \overset{\cdot\cdot}{\underset{\cdot\cdot}{Cl}}: \longrightarrow :\overset{\cdot\cdot}{\underset{\cdot\cdot}{Cl}} \overset{\cdot}{\underset{\cdot}{}} \overset{\cdot\cdot}{\underset{\cdot\cdot}{Cl}}: \longrightarrow :\overset{\cdot\cdot}{\underset{\cdot\cdot}{Cl}} - \overset{\cdot\cdot}{\underset{\cdot\cdot}{Cl}}:$$

Here each chlorine atom with seven valence electrons, contributes one electron to the shared pair. In the process of bond formation both the chlorine atoms acquire the electronic configuration of argon. In the same way, the formation of oxygen molecule involves sharing of two pairs of electrons between the two oxygen atoms. In this case both the atoms contribute two electrons each and acquire eight electrons or an octet in their valence shell.

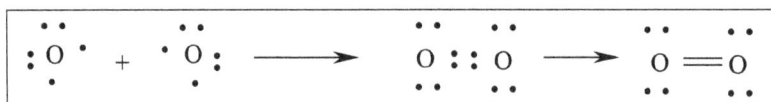

$$:\overset{\cdot\cdot}{\underset{\cdot}{O}} \cdot \; + \; \cdot \overset{\cdot\cdot}{\underset{\cdot}{O}}: \longrightarrow \overset{\cdot\cdot}{O} :: \overset{\cdot\cdot}{\underset{\cdot\cdot}{O}} \longrightarrow \overset{\cdot\cdot}{O} = \overset{\cdot\cdot}{\underset{\cdot\cdot}{O}}$$

You may have noticed that in the process of bond formation, the elements of second period acquire eight electrons in their valence shell. This is called 'Octet rule'. You may also note that in case of H_2 and Cl_2 the atoms are linked by a single line while in case of O_2 the atoms are linked by two lines. These lines represent bonds. When two atoms are bound by sharing a single pair of electron, they are said to be joined by a single bond. And when, two pairs of electrons are shared (as in case of O_2), the two atoms are said to be bound by a double bond. In nitrogen (N_2) the two atoms are joined by a triple bond as they share three pairs of electrons.

In a Lewis representation the electrons shown to be involved in the bond formation are called bonding electrons; the pair of electrons is called 'bond pair' and the pairs of electrons not involved in the bonding process are called 'lone pairs'. The nature of the electron pair plays an important role in determining the shapes of the molecules.

Polar Covalent Bond

In a chemical bond the shared electron pair is attracted by the nuclei of both the atoms. When we write the electron dot formula for a given molecule this shared electron pair is generally shown in the middle of the two atoms indicating that the two atoms attract it equally. However, actually different kinds of atoms exert different degrees of attraction on the shared pair of electrons. A more electronegative atom has greater attraction for the shared pair of electrons in a molecule. As a consequence in most cases the sharing is not equal and the shared electron pair lies more towards the atom with a higher electronegativity. For example, in HCl, the shared pair of electron is attracted more towards more electronegative chlorine atom. As a result of this unequal sharing of the electron pair, the bond acquires polarity or partial ionic character.

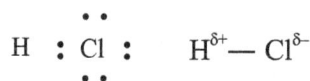

$$H \; \textbf{:} \; \overset{\displaystyle\bullet\bullet}{\underset{\displaystyle\bullet\bullet}{Cl}} \; \textbf{:} \qquad H^{\delta+}\!-\!Cl^{\delta-}$$

In an extreme case, the difference in the electronegativity may be so high that the electron pair is practically under the influence of a single atom. In other words the polarization of the bond is complete i.e., we have a case of ionic bonding. Thus, though the Lewis theory talks about covalent bonding it can account for the formation of ionic compounds also.

Coordinate Covalent Bond

You have learnt that in the formation of a covalent bond between the atoms, each atom contributes one electron to the shared electron pair, However, in some cases both the electrons of the shared pair are contributed by only one species (atom, molecule or ion) A common example is the formation of a bond between boron trifluoride (BF_3) and ammonia (NH_3). BF_3 is an electron deficient molecule and can accept a pair of electrons. The molecule of ammonia on the other hand is electron rich. It has a lone pair of electron on the nitrogen atom and that can be donated. Electron rich ammonia donates a pair of electron to electron deficient BF3.Such electron donor-acceptor bonds are called coordinate covalent or dative bonds.

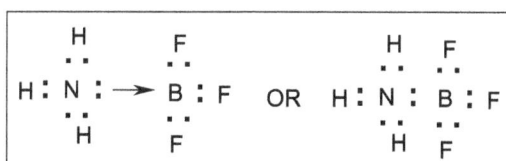

A coordinate bond is normally represented by an arrow pointing from a donor atom to the acceptor atom. A coordinate bond is identical to a covalent bond in terms of its polarity and strength. The two are different only in the way they are formed. We cannot distinguish between covalent and coordinate covalent bond, once these are formed. HNO_3 and NH_4^+ ion are some more common examples of formation of a coordinate bond.

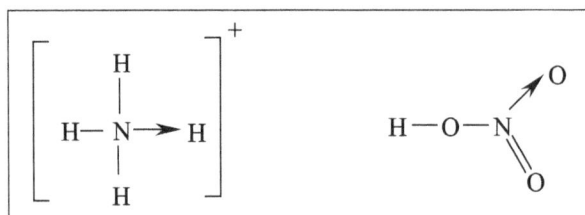

Characteristic Properties of Covalent Compounds

- The covalent compounds have low melting and boiling points due to weak forces of inter-action between the molecules.

- The covalent compounds are poor conductors of electricity as these lack ionic species.

- The covalent compounds are generally insoluble in water and dissolve in nonpolar solvents like benzene, carbon tetrachloride etc.

Hydrogen Bonding

It is a special type of attraction between a hydrogen atom bonded to a strongly electronegative atom (like nitrogen, oxygen or fluorine) and the unshared pair of electrons on another electro-negative atom. Hydrogen bond is a weak bond, the strength being just about 4-25 kJ mol^{-1}. It is quite small as compared to the covalent bond, which needs a few hundreds of kJ mol^{-1} of energy to break. However, it is strong enough to be responsible for the high boiling points of H_2O and HF etc. In fact it is due to hydrogen bonding only that water exists as a liquid. The low density of ice also can be explained in terms of hydrogen bonding.

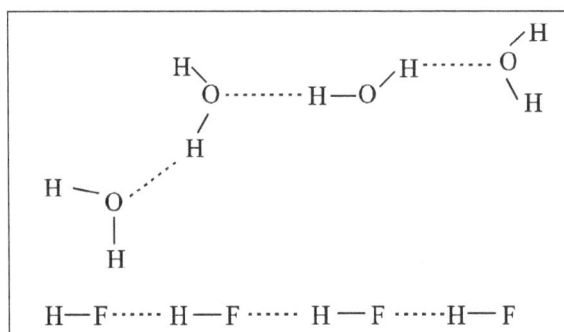

Due to the difference in the electronegativity between hydrogen and the other electronegative atom, the bond connecting them becomes polar. The hydrogen atom acquires a positive charge while the electronegative atom bears the negative charge. Hydrogen bonding results from the elec-trostatic interaction between the positively charged hydrogen atom and the negatively charged electronegative atom. The second electronegative atom may be a part of the same molecule or it may belong to a different molecule. Accordingly, there are two types of hydrogen bonds. If the hydrogen bond is formed between two different molecules it is called intermolecular hydrogen bond. When the hydrogen bond exists within the same molecule, it is called intramolecular hydrogen bonding. Salicyldehyde ad o-nitrophenol are two common examples of the molecules showing intramolecular hydrogen bonding whereas in water, intermolecular hydrogen bonding exists.

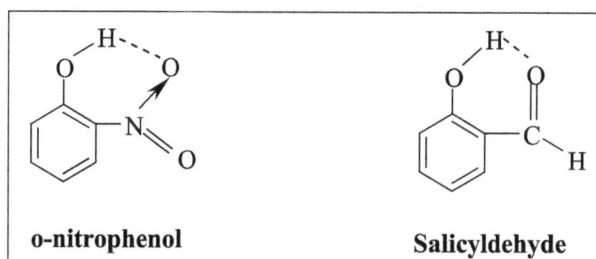

o-nitrophenol   Salicyldehyde

Hydrogen bonding plays an important role in the structure and function of many biomolecules like proteins and nucleic acids.

Valence Shell Electron Pair Repulsion (VSEPR) Theory

In a molecule the constituent atoms have definite positions relative to one another i.e., the molecules have a definite shape. The theories of bonding that we have discussed so far do not say anything about the shape of the molecules. A simple theory called VSEPR theory was put forth by Sidgwick and Powell in 1940 to explain the shapes of molecules. It was later refined and extended by Nyholm and Gillespie in1957. This theory focuses on the electron pairs present in the valence shell of the central atom of the molecule and can be stated in terms of two postulates:

Postulate 1

The electron pairs (both bonding and non-bonding) around the central atom in a molecule arrange themselves in space in such a way that they minimize their mutual repulsion. In other words, the chemical bonds in the molecule will be energetically most stable when they are as far apart from each other as possible. Let us take up some examples.

$BeCl_2$ is one of the simple triatomic molecules. In this molecule, the central atom, beryllium has an electronic configuration of 1s2 2s2. That is it has two electrons in its valence shell. In the process of covalent bond formation with two chlorine atoms two more electrons are contributed (one by each chlorine atom) to the valence shell. Thus there are a total of 4 valence electrons or two pairs of valence electrons. According to the postulate given above, these electron pairs would try to keep as far away as possible. It makes the two electron pairs to be at an angle of 180o which gives the molecule a linear shape.

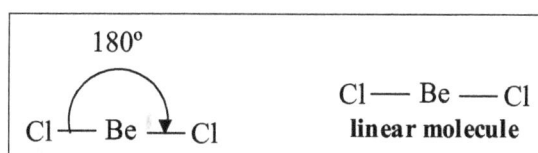

Other molecules of this type would also have a similar shape.

BF_3: In boron trifluoride, the central atom, boron has an electronic configuration of 1s^2 2s^2 2p^1. That is, it has three electrons in its valence shell. In the process of covalent bond formation with three fluorine atoms three more electrons are contributed (one by each fluorine atom) to the valence shell. Thus there are a total of 6 valence electrons or three pairs of valence electrons. According to the VSEPR postulate, these electron pairs would try to keep as far apart as possible. It makes the three electron pairs to be located at an angle of 120° which gives the molecule a planar trigonal shape.

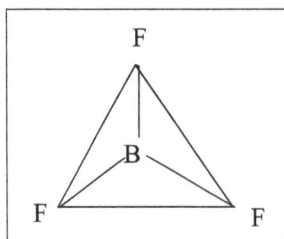

Planar trigonal shape.

Thus different molecules would have different shapes depending on the number of valence shell electrons involved. The geometric shapes associated with various numbers of electron pairs surrounding the central atom are given in table.

Table: Geometric arrangements of electron pairs around central atom.

Molecule Type	Number of electron pairs	Predicted geometry	Representative structure	Examples
AX_2	2	Linear		$HgCl_2$, BeH_2
AX_3	3	Planer trigonal		BF_3, BCl_3
AX_4	4	Tetrahedral		CCl_4, CH_4, $SiCl_4$
AX_5	5	Trigonal bipyramidal		PCl_5, PF_5
AX_6	6	Octahedral		SF_6, PF_6^-

Postulate 2

The repulsion of a lone pair of electrons for another lone pair is greater than that between a bond pair and a lone pair which in turn is greater than that between two bond pairs The order of repulsive force between different possibilities is as under.

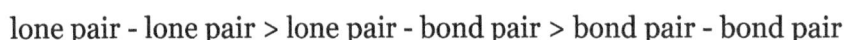

lone pair - lone pair > lone pair - bond pair > bond pair - bond pair

The shapes of the molecules given in table correspond to the molecules containing only bond pair electrons. The shapes of molecules containing a combination of lone pairs and bond pairs would be distorted from the above mentioned shapes.

Let us take an example of three molecules namely, methane, ammonia and water. All the three contain a total of 4 electron pairs around their central atom. But the nature of these is different in the three cases. In methane molecule the central carbon atom has 4 valence electrons and it shares 4 electrons with four hydrogen atoms. So there are a total of 4 bond pairs and it should have a tetrahedral shape. In case of ammonia also there are four pairs of electrons but their nature is different. Three of these are bond pairs while one is a lone pair. Similarly, in case of water again there are four pairs of electrons ; two are bond pairs while two are lone pairs. Due to the differences in the mutual repulsion between bond pair - bond pair and lone pair - bond pair the molecular shape would be slightly distorted from the expected tetrahedral shape. The number and nature of electron pairs and the geometries of these three molecules.

Molecular geometries of molecules with 4 electron pairs with different combinations of lone pairs and bond pairs.

Molecule	Number of bond pairs	Number of lone pairs	Molecular geometry	Molecular Shape	Bond angle (in degrees)
CH_4	4	0	tetrahedral		109.5
NH_3	3	1	trigonal pyramidal		107
H_2O	2	2	angular or bent		104.5

We have so far learnt that a chemical bond formation between two atoms can occur by transfer (ionic bonding) or sharing (covalent bonding) of electrons. The processes of bond formation and the bonding in simple molecules can be conveniently represented in terms of electron – dot structures. Further, the VSEPR theory provides a good idea of the shapes of the molecules. But! have you noticed that we have been representing electrons as well defined dots i.e., localized particles. This is in contradiction with the probabilistic (orbital) representation of the electron. Let us learn how do we explain the process of bond formation in terms of modern theories that incorporate the wave mechanical representation of atom.

Modern Theories of Chemical Bonding

The theories of chemical bonding proposed (in 1916) by Kossel and Lewis are called as classical theories of bonding. These do not take into account the wave mechanical or quantum mechanical principles. After the development of quantum mechanical description of atomic structure two more theories were proposed to explain the bonding between atoms. These are called modern theories of chemical bonding. These are Valence Bond Theory (VBT) and Molecular Orbital Theory (MOT).

Valence Bond Theory

Valence bond theory was proposed by Heitler and London in 1927, to describe the formation of hydrogen molecule from its atoms. Linus Pauling and others further developed it. In this approach the process of chemical bond formation can be visualised as the overlapping of atomic orbitals of the two atoms as they approach each other. The strength of the bond depends on the effectiveness or extent of the overlapping. Greater the overlapping of the orbitals, stronger is the bond formed. Let us take the example of bonding in hydrogen molecule to understand the VB approach.

Suppose that the two hydrogen atoms are at infinite distance from each other. Their electrons are in their respective 1s orbitals and are under the influence of the corresponding nuclei. As the two atoms approach each other their 1s orbitals begin to overlap which lead to decrease in energy, figure. At a distance equal to the bond length the overlapping is maximum and the energy is minimum. The overlapping can be equated to the sharing of electrons between the atoms. The electrons occupying the shared region of orbitals are under the influence of both the nuclei.

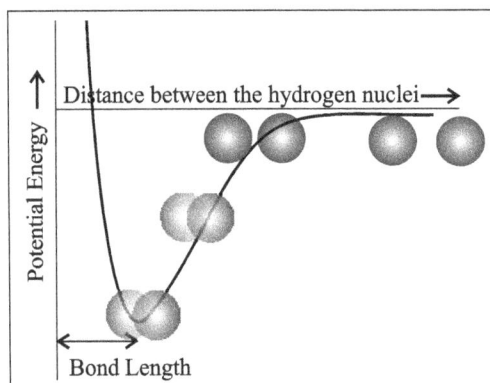

Formation of hydrogen molecule from overlapping of two hydrogen atoms.

This simple approach can be used to explain the bonding in simple diatomic molecules like HF, F_2 etc. However, to explain bonding in molecules containing more than two atoms some additional concepts like excitation and hybridisation need to be used.

Hybridisation

Let us take up the example of bonding in a triatomic molecule; say beryllium hydride (BeH2) to understand the concept of hybridisation of orbitals and the need for the same. The atomic number of beryllium is 4. Its electronic configuration is $1s^2\ 2s^2$. In order to form bonds with two hydrogen atoms the valence electrons ($2s^2$) of beryllium atom must overlap with the 1s electrons of the two hydrogen atoms. Since the valence shell of beryllium atom contains both the electrons in the same orbital (i.e., 2s) it cannot overlap with the 1s orbital of hydrogen atoms containing one electron. [You know that an orbital can contain a maximum of two electrons with opposite spin]. Pauling got over this problem by suggesting that in the process of bond formation an electron from the 2s orbital of beryllium atom gets momentarily excited to the empty 2p orbital as shown in figure.

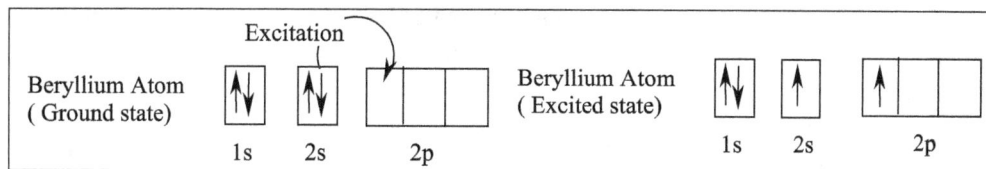

Now the two valence electrons are in two singly occupied orbitals which can overlap with the 1s orbitals of the two hydrogen atoms and form two bonds. The problem is still not over. The two bonds formed by these overlaps would be of different nature. One of these would involve overlapping of 2s orbital of beryllium with 1s orbital of hydrogen while the other would involve overlapping of 2p orbital of beryllium with 1s orbital of hydrogen. However, experimentally the two bonds are found to be equivalent.

This problem is solved with the help of a concept called hybridisation of orbitals. According to this two or more than two non equivalent orbitals (having different energies and shapes) of comparable energies mix or hybridize and give rise to an equal number of equivalent (same energies and shapes) hybrid orbitals.

In case of $BeCl_2$ the two singly occupied orbitals (2s and 2p) hybridize to give two sphybrid orbitals. This is called sp- hybridisation. These hybrid orbitals lie along the z- direction and point in opposite directions.

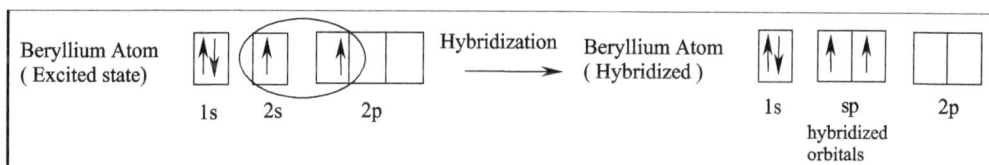

These hybrid orbitals can now overlap with the 1s orbitals of hydrogen atoms to give the linear molecule of $BeCl_2$ as shown below:

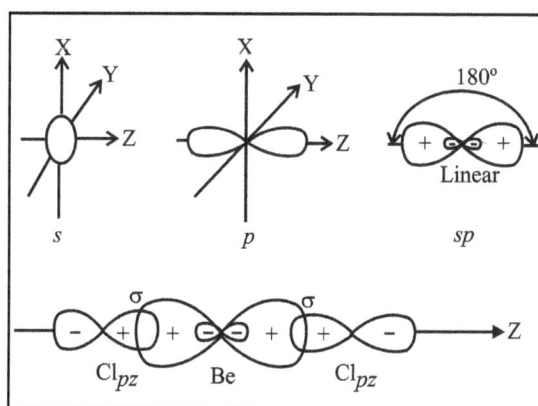

Formation of $BeCl_2$; sp hybridisation.

The concept of hybridisation as illustrated above can be used to describe the bonding and shapes of different molecules by considering hybridisation of suitable orbitals. Let us take up some more cases involving hybridisation of s and p orbitals.

Boron trichloride (sp^2 hybridisation): In boron there are five electrons and the electronic configuration is $1s^2$, $2s^2, 2p^1$. There are three electrons in the valence shell of boron atom. In order to form bonds with three chlorine atoms one of the electrons from the 2s orbital of boron atom is excited to its 2p orbital.

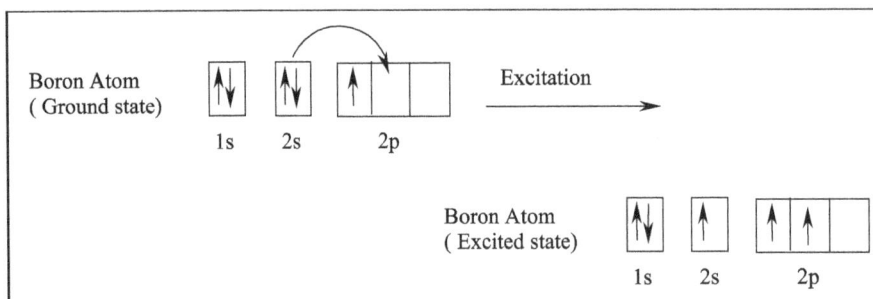

One 2s orbital and two 2p orbitals hybridise to give three sp^2 hybridized orbitals. This is called sp^2 -hybridisation.

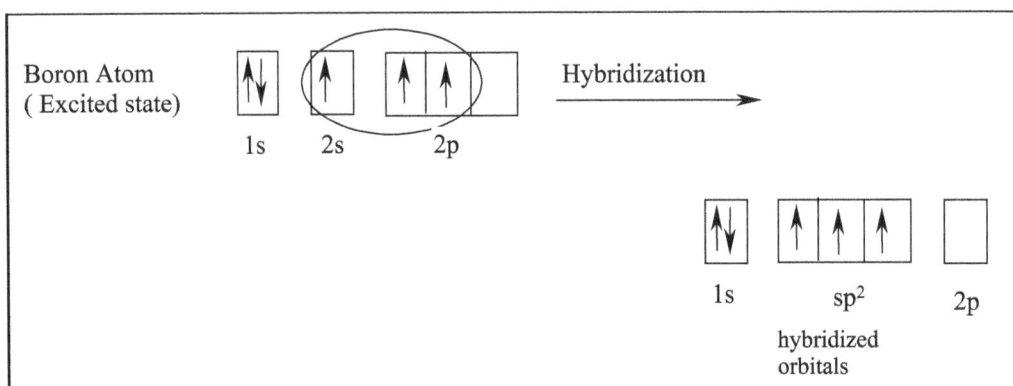

The three hybridized orbitals are coplanar and directed towards the corners of an equilateral hybrid orbitals then form bonds with the p –orbitals of chlorine atoms as shown below:

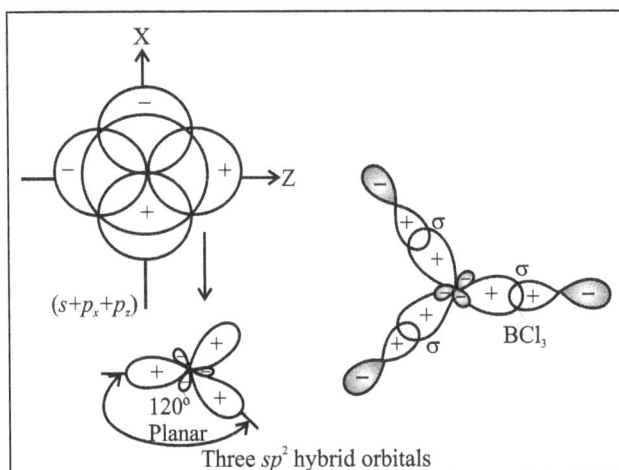

Formation of BCl3 ; sp2 hybridisation.

Bonding in Methane (sp^3 hybridisation) : In case of methane the central atom, carbon, has an electronic configuration of $1s^2$, $2s^2, 2p^2$. In order to form bonds with four hydrogen atoms one of the electrons from the 2s orbital of carbon atom is excited to the 2p orbital.

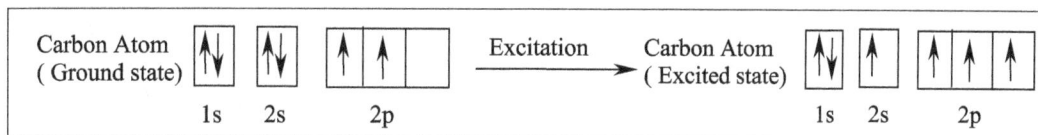

One 2s orbital and three 2p orbitals of the carbon atom then hybridize to give four sp3 hybridized orbitals. This is called sp^3- hybridisation.

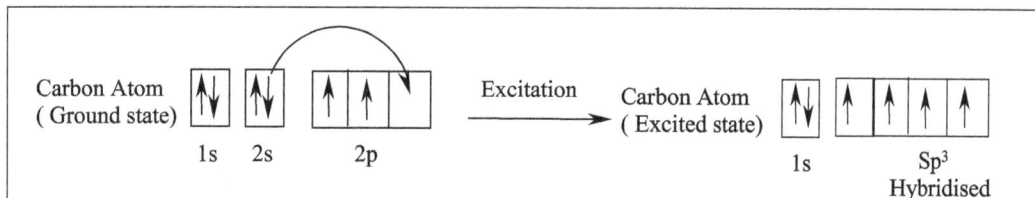

These four sp^3 hybrid orbitals are directed towards the corners of a regular tetrahedron. These hybrid orbitals then form bonds with the 1s orbitals of hydrogen atoms to give a methane molecule as shown below:

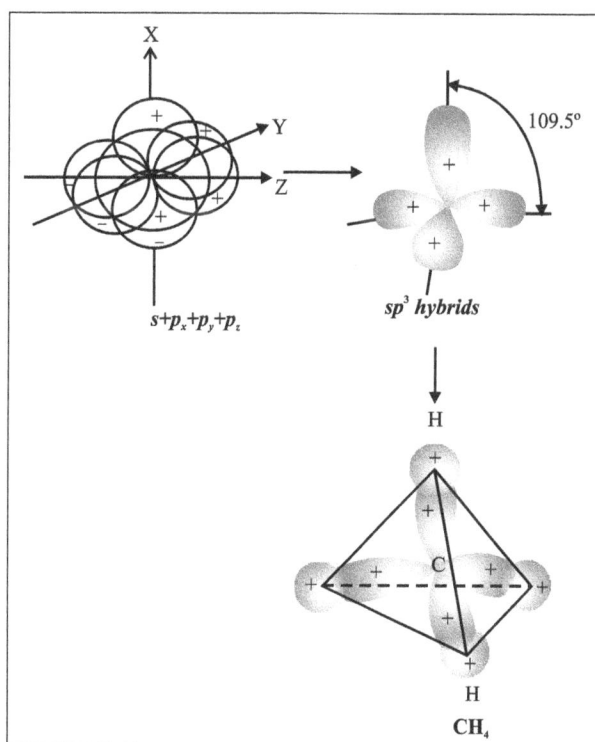

Formation of CH$_4$; sp^3 hybridisation.

Phosphorus Pentachloride (Sp^{3}d Hybridisation):

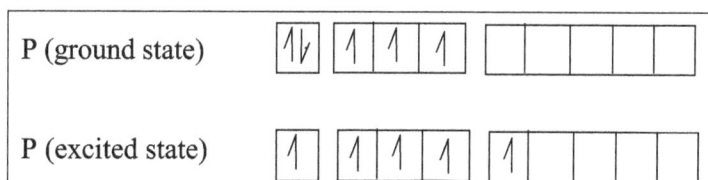

sp^{3}d hybridisation.

Five sp³d hybrid orbitals are formed which are directed towards the corners of a trigonal bipyramidal. These orbitals overlap with singly filled p-orbitals of five chlorine atoms and five σ bonds are formed. Thus PCl5 molecule has a trigonal bipyramidal geometry. Three P–Cl bonds (equatorial) make an angle of 120° with each other and lie in one plane. The other two P–Cl bonds (axial) are at 90° to the equatorial plane, one lying above and the other lying below the plane.

SF_6 (sp³d² hybridisation):

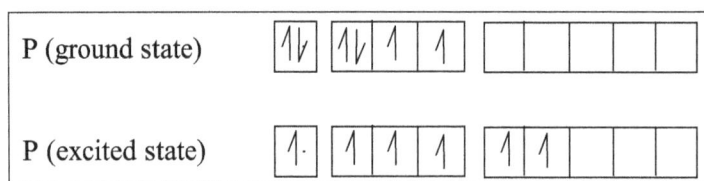

sp³d² hybridisation.

Six sp³d² hybrid orbitals are formed which are directed towards the corners of a regular octahedron. These orbitals overlap with singly filled orbitals of six F atoms and form σ bonds giving a regular octahedral geometry.

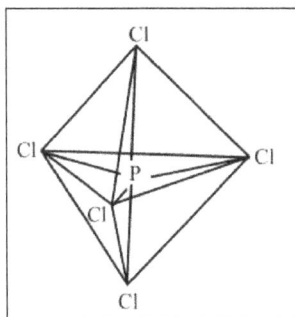

Trigonal bipyramidal geometry.

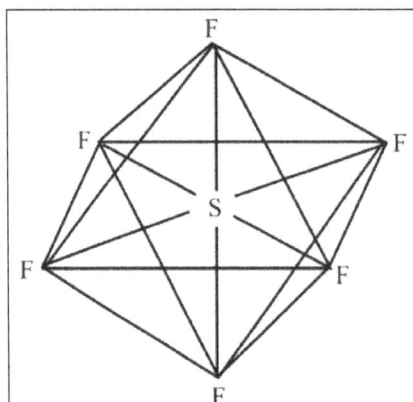

Octahedral geometry of SF6 of PCl5 molecule. molecule.

Hybridisation and Multiple Bonds

So far we have discussed the bonding in those molecules in which the orbitals on a single central atom are hybridized. Let us see how does the concept of hybridisation help us in understanding bonding between pairs of such atoms. In the case of bonding in ethane C_2H_6 two carbon atoms are bonded to each other and each carbon atom is bonded to three hydrogen atoms. You would recall that in the case of methane the valence orbitals of carbon atom undergo sp³ hybridisation. In ethane each carbon atom undergoes sp3 hybridisation to give four sp³ hybridized orbitals. The two carbon atoms form a carbon – carbon bond by sp³ - sp³ overlapping. The remaining six sp³ hybridized orbitals overlap with 1s orbitals of hydrogen atoms to give a molecule of ethane, C_2H_6.as shown in figure. The C-C bond so formed is along the internuclear axis. Such a bond is called a σ bond.

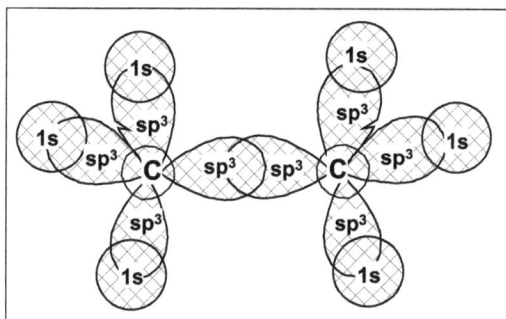

Formation of ethane molecule.

Bonding in ethene: In case of ethene, the relevant orbitals of the carbon atoms undergo sp² hybridisation. Here, only two of the three p orbitals of the carbon atoms hybridize with the 2s orbital to form three sp² hybrid orbitals each. The remaining p-orbitals (one on each carbon atom) do not take part in hybridisation. A carbon – carbon bond is formed by overlapping of sp² orbital on the two carbon atoms. The remaining four sp² hybridized orbitals overlap with the 1s orbitals of hydrogen atoms to give the basic skeleton of the molecule. This leaves an un-hybridized p orbital each on both the carbon atoms. These are perpendicular to the molecular plane and undergo sideways overlap to give an electron cloud in the plane above and below the molecule. This is called a π- bond. In ethene there are two bonds between the carbon atoms (one sigma and one pi bond).

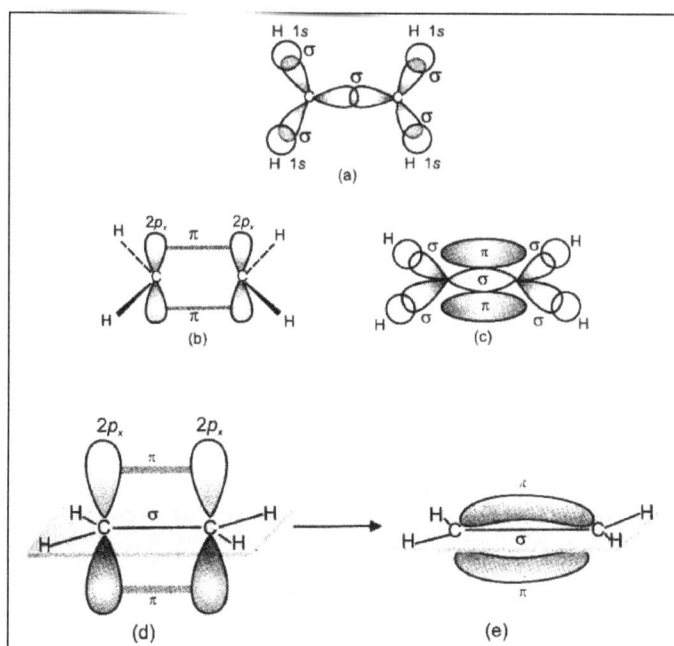

Formation of ethylene molecule: a) formation of the basic skeleton of the molecule b) sideways overlapping of the un-hybridized p orbitals and c) a π- bond (d) and (e) complete picture of ethylene molecule.

Bonding in ethyne (acetylene): In case of acetylene the bonding can be explained in terms of sp-hybridisation in carbon atoms. One 2s and one 2p orbitals hybridize to give two sp-hybridized orbitals. This leaves two mutually perpendicular unhybridised p orbitals each on both the carbon atoms. The carbon – carbon bond is formed by sp - sp overlapping with each other. The remaining sp orbital on each carbon overlaps with the 1s orbital of hydrogen to give C-H bonds. The unhybridised p orbital each on both the carbon atoms overlap sideways to give two π-bonds.

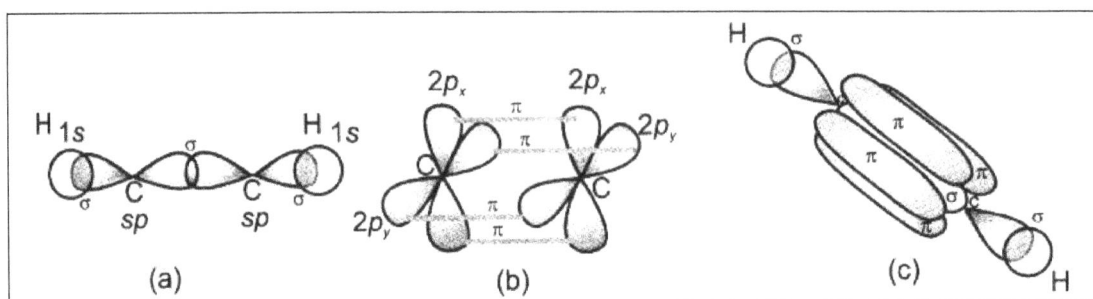

Formation of acetylene molecule : a) formation of the basic skeleton of the molecule b) sideways overlapping of two pairs of un-hybridized p orbitals and c) two mutually perpendicular π- bonds.

Molecular Orbital Theory

You have just learnt about valence bond theory. It describes bond formation as a result of overlapping of the atomic orbitals belonging to the constituent atoms. The overlapping region responsible for bonding is situated between the two atoms i.e., it is localised. Molecular orbital theory(MOT) developed by F.Hund and R.S.Mulliken in 1932, is based on the wave mechanical model of atom. In contrast to the localized bonding in VBT, the molecular orbital theory visualises the bonding to be delocalised in nature i.e., spread over the whole molecule. According to MOT, in the process of bond formation:

- The atomic orbitals of the constituent atoms combine to generate new types of orbitals (called molecular orbitals). These are spread over the whole molecule i.e., they are delocalised. In other words these new orbitals, do not "belong" to any one atom but extend over the entire region of the bonded atoms.

- These molecular orbitals are created by Linear Combination of Atomic Orbitals (LCAO) approach in which, the atomic orbitals of comparable energies and of suitable symmetry combine to give rise to an equal number of molecular orbitals.

- The available electrons then fill these orbitals in the order of increasing energy as in the Aufbau principle used in the electron configurations of atoms.

Let us take the example of hydrogen molecule to understand the molecular orbital approach to chemical bonding. The two hydrogen atoms have an electron each in their respective 1s orbitals. In the process of bond formation the atomic orbitals of two hydrogen atoms can combine in two possible ways. In one, the MO wavefunction is obtained by addition of the two atomic wave functions whereas in the other the MO is obtained by subtraction of the atomic orbitals. The combination of the 1s orbitals on the two hydrogen atoms are shown.

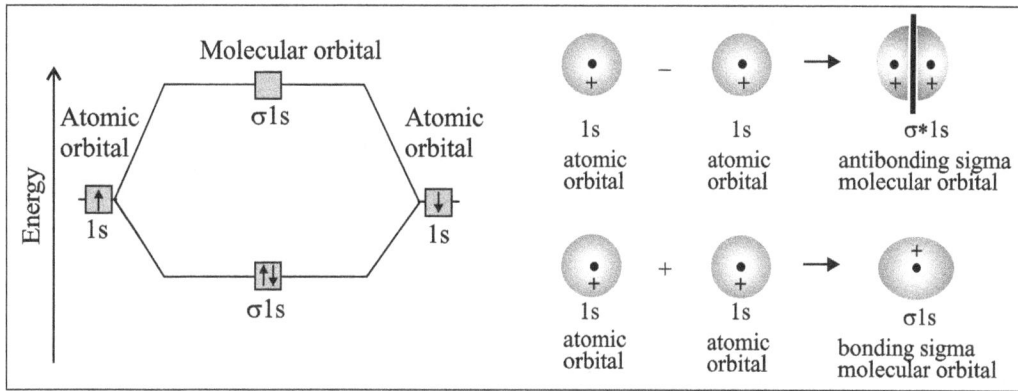

Formation of bonding (σ) and anti bonding (σ*) molecular orbitals.

The molecular orbital obtained by the addition of atomic orbitals is of lower energy than that of the atomic orbitals and is called a bonding orbital. On the other hand, the orbital obtained by subtraction of atomic orbitals is of higher energy and is called ananti-bonding orbital. You can note here that the molecular orbitals obtained here are symmetric around the bond axis (the line joining the two nuclei). Such molecular orbitals are called sigma (σ) molecular orbitals. The bonding orbital obtained above is denoted as σ1s while the anti- bonding orbital is denoted as σ*1s. Here σ indicates the type of molecular orbital; 1s tells about the atomic orbital involved and * is indicative of the anti-bonding nature of the MO. There are a total of 2 electrons in a hydrogen molecule, according to Aufbau principle these are filled intoσ1s orbital. Since the σ1sorbital is a bonding orbital, its getting filled leads to stability or the bond formation.

Like electronic configuration of atoms we write MO electronic configuration for molecules. The MO configuration of hydrogen molecule is given as $(\sigma 1s)^2$. The molecular orbital energy level diagram are given in figures.

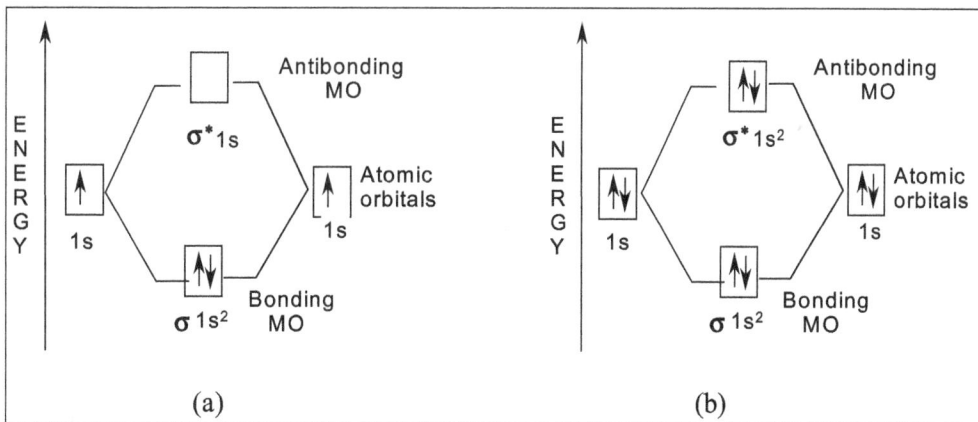

Molecular orbital energy level diagram for a) H_2 and b) He_2 molecules.

Bond Order: we may define a new parameter called bond order as,

$$\text{Bond order} = (\text{b.o.}) = \frac{1}{2}(n_b - n_a)$$

Where, n_b and n_a refer to the number of electrons present in bonding and antibonding molecular orbitals respectively. For hydrogen molecule the bond order will be ½(2−0) = 1, i.e., there is a single bond between two hydrogen atoms.

Helium (He^2) Molecule

In case of He^2, also there will be linear combination of 1s orbitals leading to the formation of σ1s and σ*1s orbitals. The four electrons would be distributed as per the MO electronic configuration : (σ1s) 2 (σ*1s) 2. The molecular orbital energy level diagram is given in figure. This gives a bond order of $\frac{1}{2}(2-2) = 0$, that is there is no bond between two helium atoms. In other words He^2 molecule does not exist.

Li$_2$ and Be$_2$ Molecules

The bonding in Li$_2$ and Be$_2$ can be explained by combining the 1s and 2s orbitals to give appropriate MO's. The molecular orbital diagrams for Li$_2$ and Be$_2$ are given:

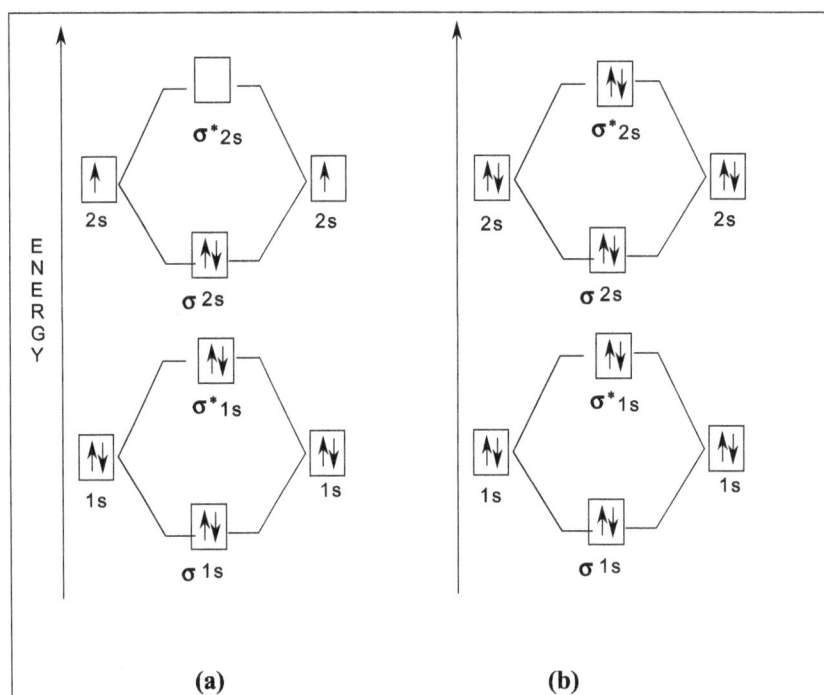

Molecular orbital energy level diagram for a) Li$_2$ and b) Be$_2$ molecules.

Molecular Orbital Bonding in Diatomic Molecules of Second Period

So far we have talked about bonding in the elements in which the MO's were obtained from the linear combination of s orbitals. In case of the atoms of second period (beyond Be) elements both s and p orbitals are involved in the formation of molecular orbitals. In such a case a number of different molecular orbitals are obtained depending on the type and symmetry of the atomic orbitals involved in the process. Let us try to understand the nature of MO's obtained in this case.

Here also the 1s and 2s orbitals of the two atoms would combine to give corresponding bonding and anti-bonding molecular orbital as shown in figure (b). Let us learn about the formation of MO's from the combination of p orbitals.

As mentioned above, in LCAO, the atomic orbitals of comparable energies and of suitable symmetry combine to give molecular orbitals. A suitable symmetry means that the combining orbitals should have same symmetry about the molecular axis. It is nomally assumed that the bond formation takes place along the z-direction. You have learnt in the first unit that the three p orbitals are directed towards three mutually perpendicular directions namely the x, y and z directions. Therefore the pz orbitals of the two atoms would combine along the bond axis to give two molecular orbitals as shown below figure. Since these molecular orbitals are symmetric around the molecular axis these are called σ orbitals. The designation of the orbitals would be σ_{2pz} and σ_{*2pz}.

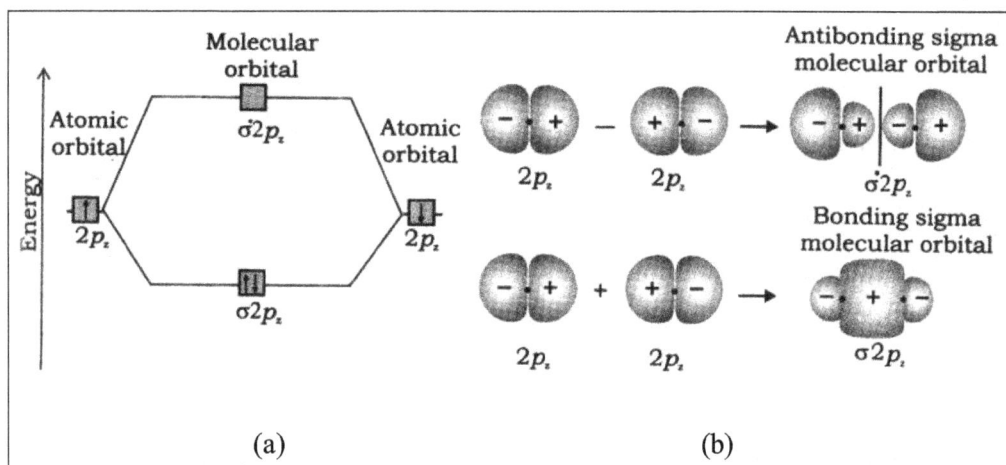

Overlapping of two $2p_z$ orbitals to give molecular orbitals.

Combination of a p_z-orbital with either a p_x or a py orbital would not lead to any bonding. On the other hand a p_x orbital will combine with a p_x and the p_y with a p_y as:

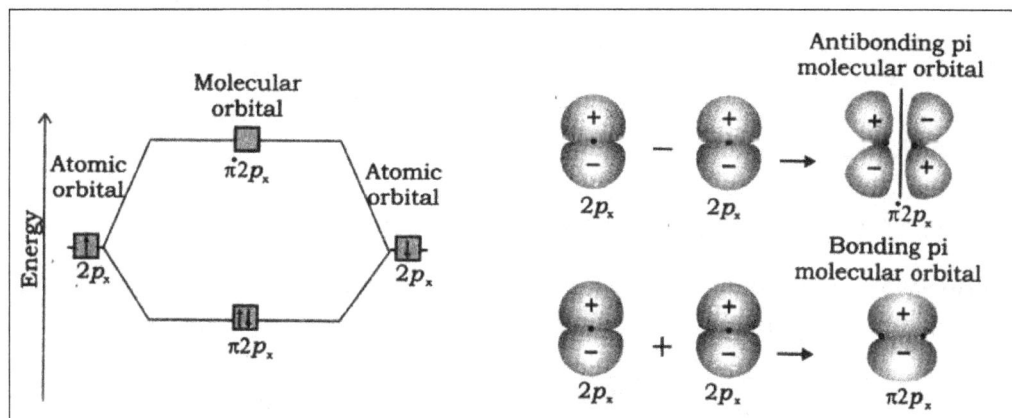

Formation of molecular orbitals from two $2p_x$ atomic orbtials.

You may note here that these orbitals combine in a lateral fashion and the resulting molecular orbitals are not symmetric around the bond axis. These MO's are called π- molecular orbitals. These have large electron density above and below the internuclear axis. The anti-bonding π orbital, π^* $2p_x$ (or π^* $2p_y$) have a node (a region of zero electron density) between the nuclei.

The molecular orbitals obtained as a result of combination of respective AO's of two atoms can be represented in the form of following energy level diagram, figure (a). The MO's obtained from the combination of 1s orbitals are not shown.(these belong to the inner core and are completely filled) The electrons in these molecular orbitals are filled in accordance with Aufbau principle and Hund's rule.

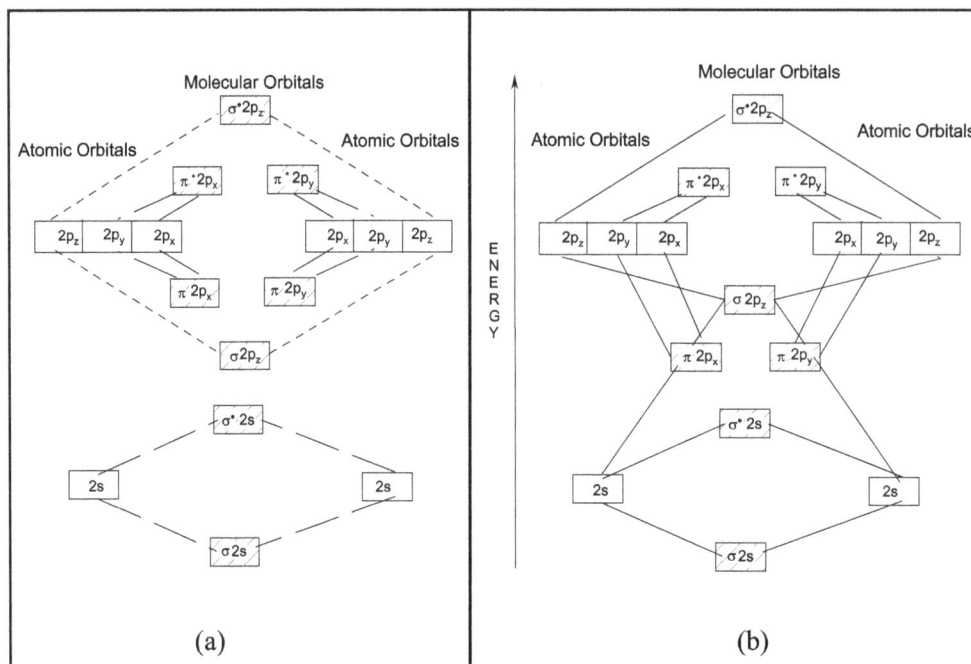

Molecular orbital energy level diagrams a) for O_2 and F_2 and b) for diatomic orbitals of lighter elements Li, Be, B, C and N.

However, this energy level diagram is valid for the diatomic molecules O_2 and F2 only; For the diatomic molecules of the lighter elements like, B, C and N this energy level diagram is somewhat modified. It is so because in case of lighter elements the difference in the energy of 2s and 2p orbitals is very low and s and p orbitals on the two atoms get mixed up. In place of normal pure 2s-2s or 2p-2p combinations we may have s-p combinations; for example 2sorbital of first atom can have a reasonable overlapping with 2pz, orbital of the second atom and vice versa. The modified energy level diagram is given in figure (b).

MO Electronic Configuration and Properties of the Molecule

The MO energy level diagram discussed above can be used to find out the MO electronic configuration of the molecule. This in turn provides the information about some properties of the molecule. Let us take the example of nitrogen molecule. An atom of nitrogen has five valence electrons; since there are two atoms, we have a total of ten valence electrons that need to be filled in the MO's. Using figure, the MO electronic configuration can be written as $\sigma 2s^2$, $\sigma *2s^2$, $\pi 2p_x^2$, $\pi 2p_y^2$, $\sigma 2p_z^2$.

Bond order: $\frac{1}{2}$ [nb−na] = $\frac{1}{2}$ [8−2] = $\frac{1}{2}$ [6]= 3; this means that in nitrogen molecule, a triple bond exists between the two nitrogen atoms.

Magnetic nature: molecules show magnetic behaviour depending on their MO electronic configuration. If all the MO's are doubly occupied the substance shows diamagnetic behaviour. In case one or more MO's are singly occupied, the substance shows paramagnetic behaviour. The MO electronic configuration of O_2 (with 12 valence electrons) is $\sigma 2s^2$, $\sigma *2s^2$, $\sigma 2p_z^2$,, $\pi 2p_x^2$, $\pi 2p_y^2$, $\pi * 2p_x^1$, $\pi * 2p_y^1$; Since it contains unpaired electrons, oxygen shows paramagnetic behaviour. This has been found to be so experimentally also. In fact, the explanation of the paramagnetic behaviour of oxygen is an achievement of MOT.

The bond order and the magnetic behaviour of the molecular cations and anions can also be obtained in the same way. In such cases we add one electron for every negative charge and for every +ve charge we subtract an electron. For example, O_2^{2-} (oxygen molecule dianion) would have a total of 14 valence electrons (12 + 2) while oxygen molecule cation O_2^+ would have 12−1 = 11 valence electrons.

References

- Atom-definition: livescience.com, Retrieved 13 March, 2019

- Molecule, definition: techtarget.com, Retrieved 25 May, 2019

- Chapter-3, english, secscicour, documents, media: nios.ac.in, Retrieved 29 April, 2019

- Measurements-2, chapter, wsu-sandbox2: lumenlearning.com, Retrieved 14 February, 2019

- 313coursee, documents, media: nios.ac.in, Retrieved 10 May, 2019

2
Branches of Chemistry

There are numerous sub-disciplines within chemistry such as electrochemistry, physical chemistry, analytical chemistry, chemical kinetics, radiochemistry and solid-state chemistry. The topics elaborated in this chapter will help in gaining a better perspective about these branches of chemistry.

Physical Chemistry

Physical chemistry is the branch of chemistry concerned with the interpretation of the phenomena of chemistry in terms of the underlying principles of physics. It lies at the interface of chemistry and physics, inasmuch as it draws on the principles of physics (especially quantum mechanics) to account for the phenomena of chemistry. It is also an essential component of the interpretation of the techniques of investigation and their findings, particularly because these techniques are becoming ever more sophisticated and because their full potential can be realized only by strong theoretical backing. Physical chemistry also has an essential role to play in the understanding of the complex processes and molecules characteristic of biological systems and modern materials.

Hot water is vaporizing as it is thrown into air that is −37.2 °C (−35 °F).

Physical chemistry is traditionally divided into a number of disciplines, but the boundaries between them are imprecise. Thermodynamics is the study of transformations of energy. Although this study might seem remote from chemistry, in fact it is vital to the study of how chemical reactions yield work and heat. Thermodynamic techniques and analyses are also used to elucidate the tendency of physical processes (such as vaporization) and chemical reactions to reach equilibrium —the condition when there is no further net tendency to change. Thermodynamics is used to relate bulk properties of substances to each other, so that measurements of one may be used to deduce the value of another. Spectroscopy is concerned with the experimental investigation of the structures of atoms and molecules, and the identification of substances, by the observation of properties of the electromagnetic radiation absorbed, emitted, or scattered by samples. Microwave spectroscopy is used to monitor the rotations of molecules; infrared spectroscopy is used to study their vibrations; and visible and ultraviolet spectroscopy is used to study electronic transitions and to infer details of electronic structures. The enormously powerful technique of nuclear magnetic resonance is now ubiquitous in chemistry. The detailed, quantitative interpretation of molecular and solid-state structure is based in quantum theory and its use in the interpretation of the nature of the chemical bond. Diffraction studies, particularly x-ray diffraction and neutron diffraction studies, provide detailed information about the shapes of molecules, and x-ray diffraction studies are central to almost the whole of molecular biology. The scattering of neutrons, in inelastic neutron scattering, gives detailed information about the motion of molecules in liquids. The bridge between thermodynamics and structural studies is called statistical thermodynamics, in which bulk properties of substances are interpreted in terms of the properties of their constituent molecules. Another major component is chemical kinetics, the study of the rates of chemical reactions; it examines, for example, how rates of reactions respond to changes in conditions or the presence of a catalyst . Chemical kinetics is also concerned with the detailed mechanisms by which a reaction takes place, the sequences of elementary processes that convert reactants into products, including chemical reactions at solid surfaces (such as electrodes).

There are further subdivisions of these major fields. Thermochemistry is a branch of thermodynamics; its focus is the heat generated or required by chemical reactions. Electrochemistry is the study of how chemical reactions can produce electricity and how electricity can drive chemical reactions in "reverse" directions (electrolysis). Increasingly, attention is shifting from equilibrium electrochemistry (which is of crucial importance in interpreting the phenomena of inorganic chemistry) to dynamic electrochemistry, in which the rates of electron-transfer processes are the focus. Chemical kinetics has divisions that are based on the rates of reaction being studied. Special techniques for studying atomic and molecular processes on ever shorter time scales are being developed, and physical chemists are now able to explore reactions on a femtosecond (10^{-15} second) timescale. Chemical kinetics studies are theoretical as well as experimental. One goal is to understand the course of reactions in step-by-step (and atomic) detail. Techniques are available that allow investigators to study collisions between individual molecules.

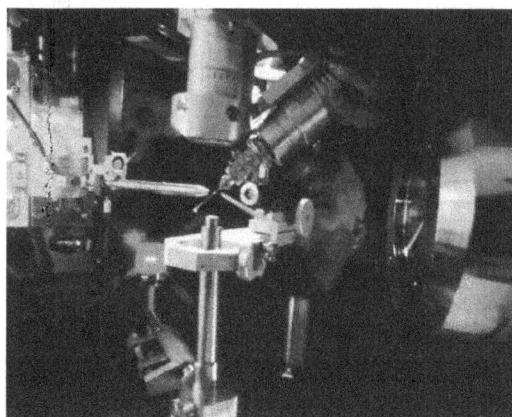

X-ray diffraction gives detailed information about shapes of molecules and is the basis of molecular biology.

Physical chemistry is essential to understanding the other branches of chemistry. It provides a basis for understanding the thermodynamic influences (principally, the entropy changes accompanying reactions) that drive chemical reactions forward. It provides justifications for the schemes proposed in organic chemistry to predict and account for the reactions of organic compounds. It accounts for the structures and properties of transition metal complexes, organometallic compounds , the microporous materials known as zeolites that are so important for catalysis , and biological macromolecules, such as proteins and nucleic acids (including DNA). It is fair to say that there is no branch of chemistry (including biochemistry) that can be fully understood without interpretations provided by physical chemistry.

There is a distinction between physical chemistry and chemical physics, although the distinction is hard to define and it is not always made. In physical chemistry, the target of investigation is typically a bulk system. In chemical physics, the target is commonly an isolated, individual molecule.

Theoretical chemistry is a branch of physical chemistry in which quantum mechanics and statistical mechanics are used to calculate properties of molecules and bulk systems. The greater part of activity in quantum chemistry, as the former is commonly termed, is the computation of the electronic structures of molecules and, often, their graphical representation. This kind of study is particularly important to the screening of compounds for potential pharmacological activity, and for establishing the mode of action of enzymes.

Analytical Chemistry

Analytical chemistry studies and uses instruments and methods used to separate, identify, and quantify matter. In practice, separation, identification or quantification may constitute the entire analysis or be combined with another method. Separation isolates analytes. Qualitative analysis identifies analytes, while quantitative analysis determines the numerical amount or concentration.

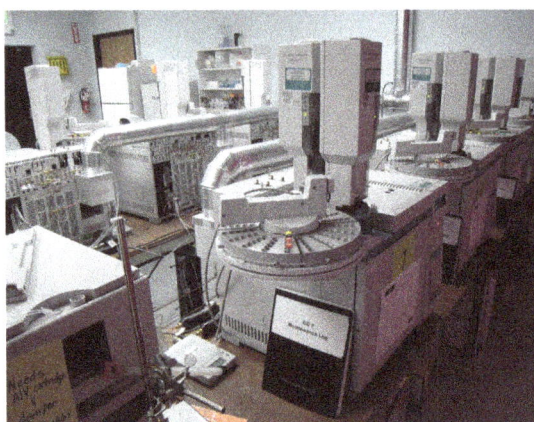

Gas chromatography laboratory.

Analytical chemistry consists of classical, wet chemical methods and modern, instrumental methods. Classical qualitative methods use separations such as precipitation, extraction, and distillation. Identification may be based on differences in color, odor, melting point, boiling point, radioactivity or reactivity. Classical quantitative analysis uses mass or volume changes to quantify amount. Instrumental methods may be used to separate samples using chromatography, electrophoresis or field flow fractionation. Then qualitative and quantitative analysis can be performed, often with the same instrument and may use light interaction, heat interaction, electric fields or magnetic fields. Often the same instrument can separate, identify and quantify an analyte.

Analytical chemistry is also focused on improvements in experimental design, chemometrics, and the creation of new measurement tools. Analytical chemistry has broad applications to forensics, medicine, science and engineering.

Classical Methods

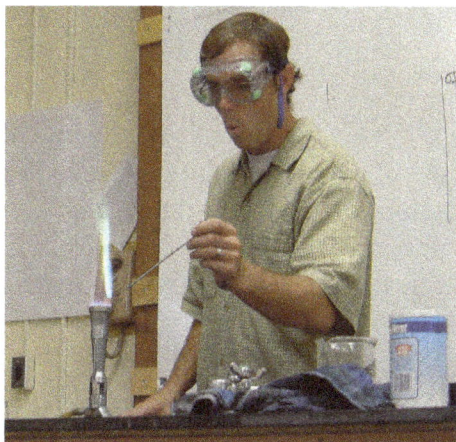

The presence of copper in this qualitative analysis is indicated by the bluish-green color of the flame.

Although modern analytical chemistry is dominated by sophisticated instrumentation, the roots of analytical chemistry and some of the principles used in modern instruments are from traditional techniques, many of which are still used today. These techniques also tend to form the backbone of most undergraduate analytical chemistry educational labs.

Qualitative Analysis

A qualitative analysis determines the presence or absence of a particular compound, but not the mass or concentration. By definition, qualitative analyses do not measure quantity.

Chemical Tests

There are numerous qualitative chemical tests, for example, the acid test for gold and the Kastle-Meyer test for the presence of blood.

Flame Test

Inorganic qualitative analysis generally refers to a systematic scheme to confirm the presence of certain aqueous ions or elements by performing a series of reactions that eliminate ranges of possibilities and then confirms suspected ions with a confirming test. Sometimes small carbon containing ions are included in such schemes. With modern instrumentation these tests are rarely used but can be useful for educational purposes and in field work or other situations where access to state-of-the-art instruments are not available or expedient.

Quantitative Analysis

Quantitative analysis is the measurement of the quantities of particular chemical constituents present in a substance.

Gravimetric Analysis

Gravimetric analysis involves determining the amount of material present by weighing the sample before and/or after some transformation. A common example used in undergraduate education is the determination of the amount of water in a hydrate by heating the sample to remove the water such that the difference in weight is due to the loss of water.

Volumetric Analysis

Titration involves the addition of a reactant to a solution being analyzed until some equivalence point is reached. Often the amount of material in the solution being analyzed may be determined. Most familiar to those who have taken chemistry during secondary education is the acid-base titration involving a color changing indicator. There are many other types of titrations, for example potentiometric titrations. These titrations may use different types of indicators to reach some equivalence point.

Instrumental Methods

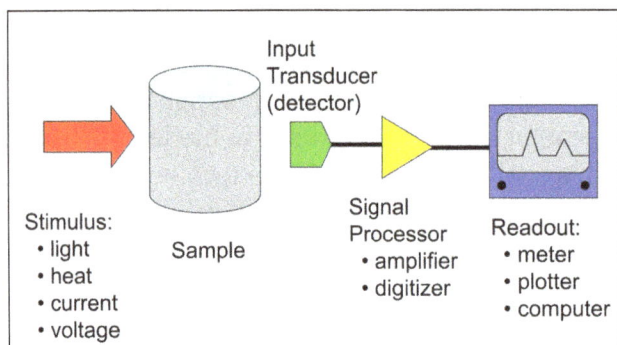

Block diagram of an analytical instrument showing the stimulus and measurement of response.

Spectroscopy

Spectroscopy measures the interaction of the molecules with electromagnetic radiation. Spectroscopy consists of many different applications such as atomic absorption spectroscopy, atomic emission spectroscopy, ultraviolet-visible spectroscopy, x-ray spectroscopy, fluorescence spectroscopy, infrared spectroscopy, Raman spectroscopy, dual polarization interferometry, nuclear magnetic resonance spectroscopy, photoemission spectroscopy, Mössbauer spectroscopy and so on.

Mass Spectrometry

Mass spectrometry measures mass-to-charge ratio of molecules using electric and magnetic fields. There are several ionization methods: electron impact, chemical ionization, electrospray, fast atom bombardment, matrix assisted laser desorption ionization, and others. Also, mass spectrometry is categorized by approaches of mass analyzers: magnetic-sector, quadrupole mass analyzer, quadrupole ion trap, time-of-flight, Fourier transform ion cyclotron resonance, and so on.

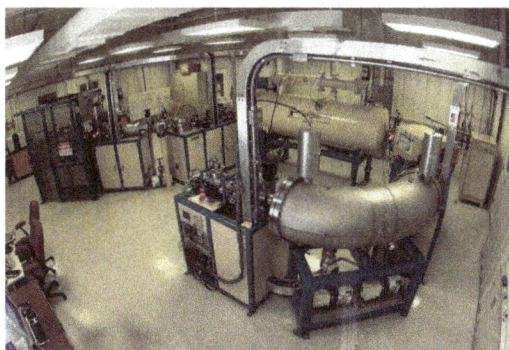

An accelerator mass spectrometer used for radiocarbon dating and other analysis.

Electrochemical Analysis

Electroanalytical methods measure the potential (volts) and/or current (amps) in an electrochemical cell containing the analyte. These methods can be categorized according to which aspects of the cell are controlled and which are measured. The four main categories are potentiometry (the difference in electrode potentials is measured), coulometry (the transferred charge is measured over time), amperometry (the cell's current is measured over time), and voltammetry (the cell's current is measured while actively altering the cell's potential).

Thermal Analysis

Calorimetry and thermogravimetric analysis measure the interaction of a material and heat.

Separation

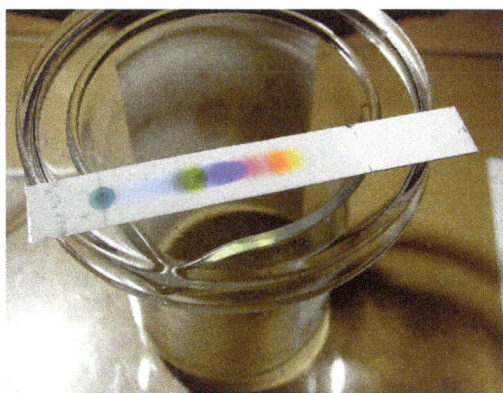

Separation of black ink on a thin-layer chromatography plate.

Separation processes are used to decrease the complexity of material mixtures. Chromatography, electrophoresis and Field Flow Fractionation are representative of this field.

Hybrid Techniques

Combinations of the above techniques produce a "hybrid" or "hyphenated" technique. Several examples are in popular use today and new hybrid techniques are under development. For example, gas chromatography-mass spectrometry, gas chromatography-infrared spectroscopy, liquid chromatography-mass spectrometry, liquid chromatography-NMR spectroscopy. liquid chromagraphy-infrared spectroscopy and capillary electrophoresis-mass spectrometry.

Hyphenated separation techniques refers to a combination of two (or more) techniques to detect and separate chemicals from solutions. Most often the other technique is some form of chromatography. Hyphenated techniques are widely used in chemistry and biochemistry. A slash is sometimes used instead of hyphen, especially if the name of one of the methods contains a hyphen itself.

Microscopy

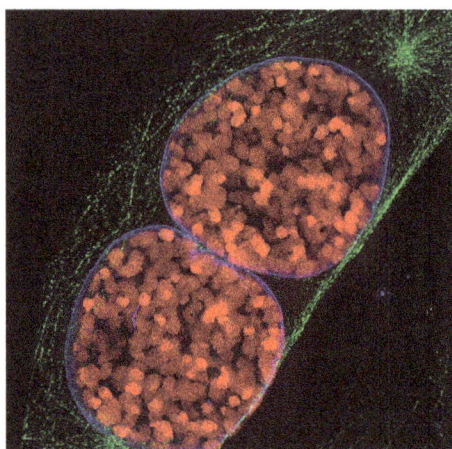

Fluorescence microscope image of two mouse cell nuclei in prophase (scale bar is 5 μm).

The visualization of single molecules, single cells, biological tissues and nanomaterials is an important and attractive approach in analytical science. Also, hybridization with other traditional analytical tools is revolutionizing analytical science. Microscopy can be categorized into three different fields: optical microscopy, electron microscopy, and scanning probe microscopy. Recently, this field is rapidly progressing because of the rapid development of the computer and camera industries.

Lab-on-a-chip

Devices that integrate (multiple) laboratory functions on a single chip of only millimeters to a few square centimeters in size and that are capable of handling extremely small fluid volumes down to less than picoliters.

Errors

Error can be defined as numerical difference between observed value and true value.

In error the true value and observed value in chemical analysis can be related with each other by the equation,

$$\varepsilon_a = | x - \bar{x} |$$

where,

ε_a is the absolute error.

x is the true value.

$\bar{x}$ is the observed value.

Error of a measurement is an inverse measure of accurate measurement i.e. smaller the error greater the accuracy of the measurement.

Errors can be expressed relatively. Given the relative error(ε_r):

$$\varepsilon_r = \frac{\varepsilon_a}{|x|} = \left|\frac{x-\bar{x}}{x}\right|$$

The percent error can also be calculated:

$$\varepsilon_r \times 100\%$$

If we want to use these values in a function, we may also want to calculate the error of the function. Let f be a function with N variables. Therefore, the propagation of uncertainty must be calculated in order to know the error in f :

$$\varepsilon_a(f) \approx \sum_{i=1}^{N}\left|\frac{\partial f}{\partial x_i}\right|\varepsilon_a(x_i) = \left|\frac{\partial f}{\partial x_1}\right|\varepsilon_a(x_1) + \left|\frac{\partial f}{\partial x_2}\right|\varepsilon_a(x_2) + \ldots + \left|\frac{\partial f}{\partial x_N}\right|\varepsilon_a(x_N)$$

Standard Curve

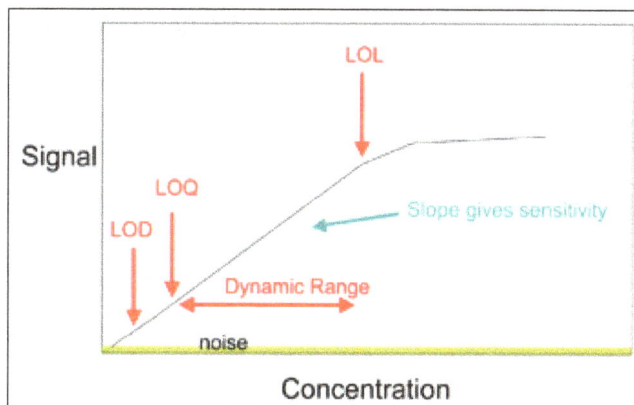

A calibration curve plot showing limit of detection (LOD), limit of quantification (LOQ), dynamic range, and limit of linearity (LOL).

A general method for analysis of concentration involves the creation of a calibration curve. This allows for determination of the amount of a chemical in a material by comparing the results of unknown sample to those of a series of known standards. If the concentration of element or compound in a sample is too high for the detection range of the technique, it can simply be diluted in a pure solvent. If the amount in the sample is below an instrument's range of measurement, the method of addition can be used. In this method a known quantity of the element or compound under study is added, and the difference between the concentration added, and the concentration observed is the amount actually in the sample.

Internal Standards

Sometimes an internal standard is added at a known concentration directly to an analytical sample to aid in quantitation. The amount of analyte present is then determined relative to the internal standard as a calibrant. An ideal internal standard is isotopically-enriched analyte which gives rise to the method of isotope dilution.

Standard Addition

The method of standard addition is used in instrumental analysis to determine concentration of a substance (analyte) in an unknown sample by comparison to a set of samples of known concentration, similar to using a calibration curve. Standard addition can be applied to most analytical techniques and is used instead of a calibration curve to solve the matrix effect problem.

Signals and Noise

One of the most important components of analytical chemistry is maximizing the desired signal while minimizing the associated noise. The analytical figure of merit is known as the signal-to-noise ratio (S/N or SNR).

Noise can arise from environmental factors as well as from fundamental physical processes.

Thermal Noise

Thermal noise results from the motion of charge carriers (usually electrons) in an electrical circuit generated by their thermal motion. Thermal noise is white noise meaning that the power spectral density is constant throughout the frequency spectrum.

The root mean square value of the thermal noise in a resistor is given by,

$$v_{RMS} = \sqrt{4k_B TR\Delta f},$$

where k_B is Boltzmann's constant, T is the temperature, R is the resistance, and Δf is the bandwidth of the frequency f.

Shot Noise

Shot noise is a type of electronic noise that occurs when the finite number of particles (such as electrons in an electronic circuit or photons in an optical device) is small enough to give rise to statistical fluctuations in a signal.

Shot noise is a Poisson process and the charge carriers that make up the current follow a Poisson distribution. The root mean square current fluctuation is given by,

$$i_{RMS} = \sqrt{2eI\Delta f}$$

where, e is the elementary charge and I is the average current. Shot noise is white noise.

Flicker Noise

Flicker noise is electronic noise with a $1/f$ frequency spectrum; as f increases, the noise decreases. Flicker noise arises from a variety of sources, such as impurities in a conductive channel, generation and recombination noise in a transistor due to base current, and so on. This noise can be avoided by modulation of the signal at a higher frequency, for example through the use of a lock-in amplifier.

Environmental Noise

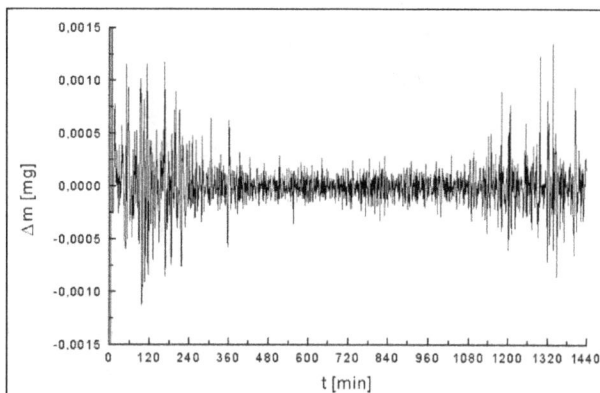

Noise in a thermogravimetric analysis; lower noise in the middle of the plot results from less human activity (and environmental noise) at night.

Environmental noise arises from the surroundings of the analytical instrument. Sources of electromagnetic noise are power lines, radio and television stations, wireless devices, Compact fluorescent lamps and electric motors. Many of these noise sources are narrow bandwidth and therefore can be avoided. Temperature and vibration isolation may be required for some instruments.

Noise Reduction

Noise reduction can be accomplished either in computer hardware or software. Examples of hardware noise reduction are the use of shielded cable, analog filtering, and signal modulation. Examples of software noise reduction are digital filtering, ensemble average, boxcar average, and correlation methods.

Applications

Analytical chemistry has applications including in forensic science, bioanalysis, clinical analysis, environmental analysis, and materials analysis. Analytical chemistry research is largely driven by performance (sensitivity, detection limit, selectivity, robustness, dynamic range, linear range, accuracy, precision, and speed), and cost (purchase, operation, training, time, and space). Among the main branches of contemporary analytical atomic spectrometry, the most widespread and universal are optical and mass spectrometry. In the direct elemental analysis of solid samples, the new leaders are laser-induced breakdown and laser ablation mass spectrometry, and the related techniques with transfer of the laser ablation products into inductively coupled plasma. Advances in design of diode lasers and optical parametric oscillators promote developments in fluorescence and ionization spectrometry and also in absorption techniques where uses of optical cavities for increased effective absorption pathlength are expected to expand. The use of plasma- and laser-based methods is increasing. An interest towards absolute (standardless) analysis has revived, particularly in emission spectrometry.

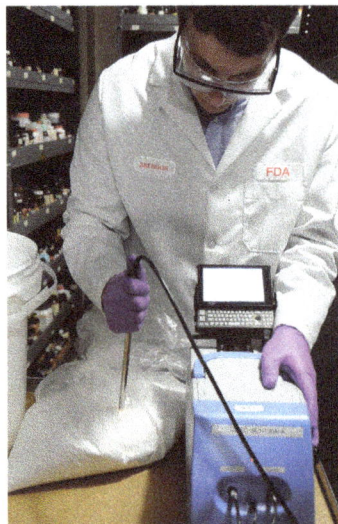

US Food and Drug Administration scientist uses portable near infrared spectroscopy device to detect potentially illegal substances.

Great effort is being put in shrinking the analysis techniques to chip size. Although there are few examples of such systems competitive with traditional analysis techniques, potential advantages include size/portability, speed, and cost. (micro total analysis system (µTAS) or lab-on-a-chip). Microscale chemistry reduces the amounts of chemicals used.

Many developments improve the analysis of biological systems. Examples of rapidly expanding fields in this area are genomics, DNA sequencing and related research in genetic fingerprinting and DNA microarray; proteomics, the analysis of protein concentrations and modifications, especially in response to various stressors, at various developmental stages, or in various parts of the body, metabolomics, which deals with metabolites; transcriptomics, including mRNA and associated fields; lipidomics - lipids and its associated fields; peptidomics - peptides and its associated fields; and metalomics, dealing with metal concentrations and especially with their binding to proteins and other molecules.

Analytical chemistry has played critical roles in the understanding of basic science to a variety of practical applications, such as biomedical applications, environmental monitoring, quality control of industrial manufacturing, forensic science and so on.

The recent developments of computer automation and information technologies have extended analytical chemistry into a number of new biological fields. For example, automated DNA sequencing machines were the basis to complete human genome projects leading to the birth of genomics. Protein identification and peptide sequencing by mass spectrometry opened a new field of proteomics.

Analytical chemistry has been an indispensable area in the development of nanotechnology. Surface characterization instruments, electron microscopes and scanning probe microscopes enables scientists to visualize atomic structures with chemical characterizations.

Solid-state Chemistry

Solid-state chemistry, also sometimes referred as materials chemistry, is the study of the synthesis, structure, and properties of solid phase materials, particularly, but not necessarily exclusively of, non-molecular solids. It therefore has a strong overlap with solid-state physics, mineralogy, crystallography, ceramics, metallurgy, thermodynamics, materials science and electronics with a focus on the synthesis of novel materials and their characterisation. Solids can be classified as crystalline or amorphous on basis of the nature of order present in the arrangement of their constituent particles.

Synthetic Methods

Given the diversity of solid state compounds, an equally diverse array of methods are used for their preparation.

Chemie Douce

For organic materials, such as charge transfer salts, the methods operates near room temperature and are often similar to the techniques of organic synthesis. Redox reactions are sometimes conducted by electrocrystallisation, as illustrated by the preparation of the Bechgaard salts from tetrathiafulvalene.

Oven Techniques

For thermally robust materials, high temperature methods are often employed. For example, bulk solids are prepared using tube furnaces, which allow reactions to be conducted up to ca. 1100 °C. Special equipment e.g. ovens consisting of a tantalum tube through which an electric current is passed can be used for even higher temperatures up to 2000 °C. Such high temperatures are at times required to induce diffusion of the reactants.

Tube furnace being used during the synthesis of aluminium chloride.

Melt Methods

One method often employed is to melt the reactants together and then later anneal the solidified melt. If volatile reactants are involved the reactants are often put in an ampoule that is evacuated mixture.

By keeping the bottom of the ampoule in liquid nitrogen- and then sealed. The sealed ampoule is then put in an oven and given a certain heat treatment.

Solution Methods

It is possible to use solvents to prepare solids by precipitation or by evaporation. At times the solvent is used hydrothermal that is under pressure at temperatures higher than the normal boiling point. A variation on this theme is the use of flux methods, where a salt of relatively low melting point is added to the mixture to act as a high temperature solvent in which the desired reaction can take place. this can be very useful.

Gas Reactions

Chemical vapour deposition reaction chamber.

Many solids react vigorously with reactive gas species like chlorine, iodine, oxygen etc. Others form adducts with other gases, e.g. CO or ethylene. Such reactions are often conducted in a tube that is open ended on both sides and through which the gas is passed. A variation of this is to let the reaction take place inside a measuring device such as a TGA. In that case stoichiometric information can be obtained during the reaction, which helps identify the products.

A special case of a gas reaction is a chemical transport reaction. These are often carried out in a sealed ampoule to which a small amount of a transport agent, e.g. iodine is added. The ampoule is then placed in a zone oven. This is essentially two tube ovens attached to each other which allows a temperature gradient to be imposed. Such a method can be used to obtain the product in the form of single crystals suitable for structure determination by X-ray diffraction.

Chemical vapour deposition is a high temperature method that is widely employed for the preparation of coatings and semiconductors from molecular precursors.

Characterization

New Phases, Phase Diagrams and Structures

Synthetic methodology and characterization often go hand in hand in the sense that not one but a series of reaction mixtures are prepared and subjected to heat treatment. The stoichiometry is typically *varied* in a systematic way to find which stoichiometries will lead to new solid compounds or to solid solutions between known ones. A prime method to characterize the reaction products is powder diffraction, because many solid state reactions will produce polycristalline ingots or powders. Powder diffraction will facilitate the identification of known phases in the mixture. If a pattern is found that is not known in the diffraction data libraries an attempt can be made to index the pattern, i.e. to identify the symmetry and the size of the unit cell. (If the product is not crystalline the characterization is typically much more difficult.)

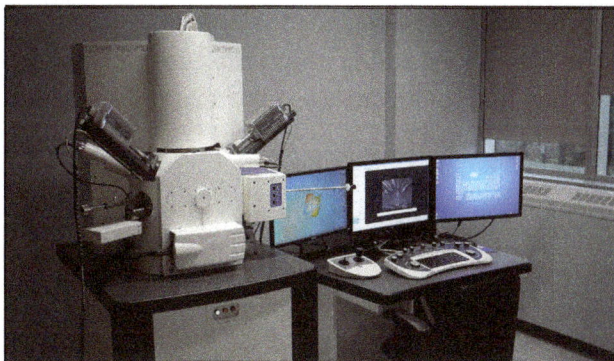

Scanning Electron Microscope (SEM).

Once the unit cell of a new phase is known, the next step is to establish the stoichiometry of the phase. This can be done in a number of ways. Sometimes the composition of the original mixture will give a clue, if one finds only one product -a single powder pattern- or if one was trying to make a phase of a certain composition by analogy to known materials but this is rare. Often considerable effort in refining the synthetic methodology is required to obtain a pure sample of the new material. If it is possible to separate the product from the rest of the reaction mixture elemental analysis can be used. Another way involves SEM and the generation of characteristic X-rays in the electron beam. X-ray diffraction is also used due to its imaging capabilities and speed of data generation.

The latter often requires *revisiting* and refining the preparative procedures and that is linked to the question which phases are stable at what composition and what stoichiometry. In other words, what does the phase diagram looks like. An important tool in establishing this is thermal analysis techniques like DSC or DTA and increasingly also, thanks to the advent of synchrotrons temperature-dependent powder diffraction. Increased knowledge of the phase relations often leads to further refinement in synthetic procedures in an iterative way.

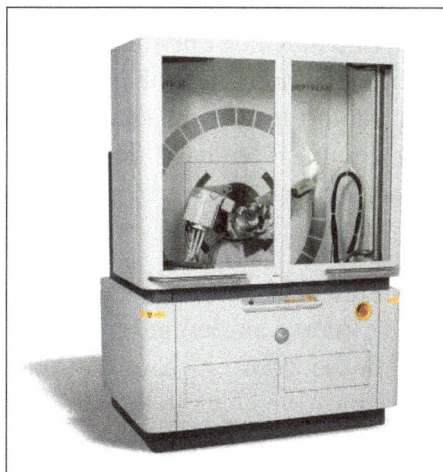

X-ray diffractometer (XRD).

New phases are thus characterized by their melting points and their stoichiometric domains. The latter is important for the many solids that are non-stoichiometric compounds. The cell parameters obtained from XRD are particularly helpful to characterize the homogeneity ranges of the latter.

Further Characterization

In many -but certainly not all- cases new solid compounds are further characterized by a variety of techniques that straddle the fine line that (hardly) separates solid-state chemistry from solid-state physics.

Optical Properties

For non-metallic materials, it is often possible to obtain UV/VIS spectra. In the case of semiconductors that will give an idea of the band gap.

Chemical Kinetics

Chemical kinetics is the branch of physical chemistry that is concerned with understanding the rates of chemical reactions. It is to be contrasted with thermodynamics, which deals with the direction in which a process occurs but in itself tells nothing about its rate. Thermodynamics is time's arrow, while chemical kinetics is time's clock. Chemical kinetics relates to many aspects of cosmology, geology, biology, engineering, and even psychology and thus has far-reaching implications. The principles of chemical kinetics apply to purely physical processes as well as to chemical reactions.

One reason for the importance of kinetics is that it provides evidence for the mechanisms of chemical processes. Besides being of intrinsic scientific interest, knowledge of reaction mechanisms is of practical use in deciding what is the most effective way of causing a reaction to occur. Many commercial processes can take place by alternative reaction paths, and knowledge of the mechanisms makes it possible to choose reaction conditions that favour one path over others.

A chemical reaction is, by definition, one in which chemical substances are transformed into other substances, which means that chemical bonds are broken and formed so that there are changes in the relative positions of atoms in molecules. At the same time, there are shifts in the arrangements of the electrons that form the chemical bonds. A description of a reaction mechanism must therefore deal with the movements and speeds of atoms and electrons. The detailed mechanism by which a chemical process occurs is referred to as the reaction path, or pathway.

The vast amount of work done in chemical kinetics has led to the conclusion that some chemical reactions go in a single step; these are known as elementary reactions. Other reactions go in more than one step and are said to be stepwise, composite, or complex. Measurements of the rates of chemical reactions over a range of conditions can show whether a reaction proceeds by one or more steps. If a reaction is stepwise, kinetic measurements provide evidence for the mechanism of the individual elementary steps. Information about reaction mechanisms is also provided by certain nonkinetic studies, but little can be known about a mechanism until its kinetics has been investigated. Even then, some doubt must always remain about a reaction mechanism. An investigation, kinetic or otherwise, can disprove a mechanism but can never establish it with absolute certainty.

Reaction Rate

The rate of a reaction is defined in terms of the rates with which the products are formed and the reactants (the reacting substances) are consumed. For chemical systems it is usual to deal with the concentrations of substances, which is defined as the amount of substance per unit volume. The rate can then be defined as the concentration of a substance that is consumed or produced in unit time. Sometimes it is more convenient to express rates as numbers of molecules formed or consumed in unit time.

The Half-life

A useful rate measure is the half-life of a reactant, which is defined as the time that it takes for half of the initial amount to undergo reaction. For a special type of kinetic behaviour, the half-life is independent of the initial amount. A common and straightforward example of a half-life independent of the initial amount is radioactive substances. Uranium-238, for example, decays with a half-life of 4.5 billion years; of an initial amount of uranium, half of that amount will have decayed in that period of time. The same behaviour is found in many chemical reactions.

Even when the half-life of a reaction varies with the initial conditions, it is often convenient to quote a half-life, bearing in mind that it applies only to the particular initial conditions. Consider, for example, the reaction in which hydrogen and oxygen gases combine to form water; the chemical equation is,

$$2H_2 + O_2 \rightarrow 2H_2O.$$

If the gases are mixed together at atmospheric pressure and room temperature, nothing observable will happen over long periods of time. However, reaction does occur, with a half-life that is estimated to be more than 12 billion years, which is roughly the age of the universe. If a spark is passed through the system, the reaction occurs with explosive violence, with a half-life of less than one-millionth of a second. This is a striking example of the great range of rates with which chemical kinetics is concerned. There are many possible processes that proceed too slowly to be studied experimentally, but sometimes they can be accelerated, often by the addition of a substance known as a catalyst. Some reactions are even faster than the hydrogen-oxygen explosion—for example, the combination of atoms or molecular fragments (called free radicals) where all that occurs is the formation of a chemical bond. Some modern kinetic investigations are concerned with even faster processes, such as the breakdown of highly energetic and therefore transient molecules, where times of the order of femtoseconds (fs; 1 fs = 10^{-15} second) are involved.

Measuring Slow Reactions

The best way to study exceedingly slow reactions is to change the conditions so that the reactions occur in a reasonable time. Increasing the temperature, which can have a strong effect on the reaction rate, is one possibility. If the temperature of a hydrogen-oxygen mixture is raised to about 500 °C (900 °F), reaction then occurs rapidly, and its kinetics has been studied under those conditions. When a reaction occurs to a measurable extent over a period of minutes, hours, or days, rate measurements are straightforward. Amounts of reactants or products are measured at various times, and the rates are readily calculated from the results. Many automated systems have now been devised for measuring rates in this way.

Measuring Fast Reactions

Some processes are so fast that special techniques have to be used to study them. There are two difficulties with fast reactions. One is that the time that it takes to mix reactants or to change the temperature of the system may be significant in comparison with the half-life, so that the initial time cannot be measured accurately. The other difficulty is that the time it takes to measure the amounts of substances may be comparable with the half-life of the reaction. The methods used to overcome these difficulties fall into two classes: flow methods and pulse and probe methods.

In flow methods, two gases or solutions are introduced rapidly into a mixing vessel, and the resulting mixture then flows rapidly along a tube. Concentrations of reactants or products may then be measured—for example, by spectroscopic methods—at various positions along the tube, which correspond to various reaction times. A modification of this method is the stopped-flow technique, in which the reactants are forced rapidly into a reaction chamber; the flow is then suddenly stopped, and the amounts are measured by physical methods after various short times. These flow methods are limited by the time it takes to mix gases or solutions and are not suitable if the half-life is less than about a hundredth of a second.

These mixing difficulties were overcome by pulse and probe methods. The principle of these is that a short pulse, usually of radiation, is given to a chemical system and is then followed by a probe, usually involving radiation that provides spectroscopic evidence of what occurred after the initial pulse. The first of these methods, developed in 1949 by British chemists R.G.W. Norrish and George Porter, was the flash-photolysis method, for which Norrish and Porter won the Nobel Prize for Chemistry in 1967. In this technique a flash of light of high intensity but short duration brings about the formation of atomic and molecular species, the reactions of which can be studied kinetically by spectroscopy. In the earliest experiments the duration of the flash was about a millisecond (ms; $1 ms = 10^{-3}$ second), but in the next four decades the duration was reduced by more than 11 powers of 10, to just a few femtoseconds. A nanosecond (ns; $1 ns = 10^{-9}$ second) flash is adequate for studying almost any purely chemical reaction where there is a change in chemical identity. Any chemical reaction, however, involves processes of a purely physical nature, such as energy redistribution and the breakdown of transient species, which occur in the femtosecond range.

Many such processes have now been studied with flashes of only a few femtoseconds' duration. The time that it takes for the length of a chemical bond to change by 10–10 metre can be as little as about 100 fs, so that a flash of a few femtoseconds' duration, closely followed by another one of the same duration, will provide information about such tiny changes in bond lengths. The technique for causing one flash to occur a few nanoseconds after another is to route the light by a slightly longer path. A path of 1 additional micrometre (μm; $1 μm = 10^{-6}$ metre) causes a delay of 1 fs, and such a short path difference is now technically feasible. Egyptian-born chemist Ahmed Zewail won the Nobel Prize for Chemistry in 1999 for his work in this field.

Another pulse method is the relaxation method, developed in the 1950s by German physicist Manfred Eigen (who shared the Nobel Prize for Chemistry in 1967 with Norrish and Porter). In this method the investigation begins with a reaction system in equilibrium; the reaction to be studied has finished, and no further changes take place. The external conditions are then altered very rapidly; the system is then no longer at equilibrium, and it relaxes to a new equilibrium. The speed of relaxation is measured by a physical method such as spectroscopy, and analysis of the results leads to the reaction rate.

The most common way of changing the external conditions is to change the temperature, and the method is called the temperature-jump, or T-jump, method. Techniques have been developed for raising the temperature of a tiny reaction vessel by a few degrees in less than 100 ns. The method is therefore not suitable for the fastest processes, which can be studied by flash photolysis, but many purely chemical processes are suitable for the T-jump technique, which has provided valuable kinetic information.

Other experimental techniques are used for the study of rapid processes. Ultrasonic methods have been used for processes occurring with half-lives in the microsecond (μs; $1\ \mu s = 10^{-6}$ second) and nanosecond ranges. Nuclear magnetic resonance has also been used for certain types of reactions.

Some Kinetic Principles

The kinetic behaviour of an ordinary chemical reaction is conventionally studied in the first instance by determining how the reaction rate is influenced by certain external factors such as the concentrations of the reacting substances, the temperature, and sometimes the pressure. For a reaction in which two substances A and B react with each other, it is sometimes found that the reaction rate is proportional to the concentration of A, represented by [A], and to the concentration of B, or [B]. In that case the reaction is said to be a second-order reaction; it is first order in [A] and first order in [B]. In such a case the reaction rate v can be expressed as,

$$v = k[A][B],$$

where k is a constant, known as the rate constant for the reaction.

This is just one of many types of kinetics that can be observed. A substance A that changes into another substance may obey a kinetic equation of the form $v = k[A]$, which is a first-order reaction. It is important to recognize that the kinetics of a reaction does not always correspond in a simple way to the balanced chemical equation for the reaction. Thus, if a reaction is of the form,

$$A + B \rightleftharpoons Y + Z,$$

the reaction is not necessarily second-order in both directions. This is in contrast to the situation with the equilibrium constant for the reaction, which corresponds to the balanced equation. The reason why the kinetic law is different is that the reactions in the forward and reverse directions may occur by stepwise mechanisms that lead to a different and usually more complex kinetic equation.

Sometimes reaction rates depend on reactant concentrations in a more complicated way. This is a clear indication that a reaction happens in several steps.

The effect of temperature on reaction rates provides much information about reaction mechanisms. Understanding of this effect owes much to the ideas of the Dutch physical chemist Jacobus Henricus van 't Hoff and the Swedish chemist Svante August Arrhenius. Their equation for the dependence of a rate constant k on the absolute temperature T is,

$$k = A \exp(-E/RT),$$

where R is the molar gas constant and A and E are quantities that are different for each reaction. This equation has come to be called the Arrhenius equation, although, as Arrhenius acknowledged when he applied it in 1889, it was first suggested by van 't Hoff in 1884. According to this relationship, a plot of the logarithm of the rate constant against the reciprocal of the absolute temperature should yield a straight line. From the slope and intercepts of the line, it is possible to calculate the value of the kinetic parameters A and E. The Arrhenius relationship applies satisfactorily to most reactions and indeed to many physical processes; however, various complications may cause it to fail.

If the reaction between two molecules is an elementary one, occurring in a single step, a simple interpretation of the Arrhenius equation can be given. The quantity A is related to the frequency of collisions between the reacting molecules. The quantity E, known as the activation energy for the reaction, results from the fact that there is an energy barrier to reaction. If E was zero, k would be equal to A, which means that the reaction would occur every time a collision occurred between the reactant molecules. This is the case for reactions in which no chemical bond is broken, such as the combination of atoms.

For reactions in which a chemical bond is broken, on the other hand, the activation energy E is not zero but has a value that is often a tenth or so of the energy required to break the bond. A simple and essentially correct explanation of the activation energy was suggested by Arrhenius, who pointed out that, for many reactions, raising the temperature by 10 °C (18 °F) doubles the reaction rate. This increase cannot be caused by the increase in the frequency of collisions between colliding molecules, since the frequency does not increase sufficiently with a rise in temperature. Arrhenius suggested that when reactants A and B react together, they first form a highly energized intermediate that is denoted as $AB^{\ddagger}$, which subsequently gives the products of reaction,

$$A + B \rightleftharpoons AB^{\ddagger} \rightarrow Y + Z.$$

If the intermediate complex (also called the activated complex) $AB^{\ddagger}$ is of high energy, it is formed only in small amounts. According to the Boltzmann principle, the fraction of molecules having energy greater than E is $\exp(-E/RT)$, which provides an explanation of the appearance of this fraction in the Arrhenius equation. The interpretation of the equation is thus that only those molecules having energy greater than E are able to undergo reaction; other collisions are ineffective, and the reactant molecules merely separate unchanged.

Composite Reaction Mechanisms

Various lines of evidence are used to determine if a reaction occurs in more than one step. Suppose that the kinetic equation for the reaction does not correspond to the balanced equation for the reaction. A simple example is the reaction between hydrogen and iodine chloride, with the formation of iodine and hydrogen chloride,

$$H_2 + 2ICl \rightarrow I_2 + 2HCl.$$

To make the equation balance, the reaction must be written as shown, with two iodine chloride molecules reacting with a single hydrogen molecule. If this reaction occurred in a single elementary step, the rate would be proportional to the first power of the hydrogen concentration and the square of the iodine chloride concentration. Instead, however, the rate is found to be proportional to both concentrations to the first power, so that it is a second-order reaction: $v = k[H_2][ICl]$. This can be explained if there is initially a slow reaction between one hydrogen molecule and one of iodine chloride:

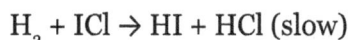

$H_2 + ICl \rightarrow HI + HCl$ (slow)

Followed by a rapid reaction between the hydrogen iodide formed and an additional molecule of iodine chloride:

$HI + ICl \rightarrow HCl + I_2$ (fast).

If the second reaction is fast, the hydrogen iodide is removed as fast as it is formed. The rate of the second reaction therefore has no effect on the overall rate, which is the rate of the first step. This mechanism therefore explains the kinetic behaviour but does not prove it; other, more complicated schemes could be devised, but, until there is further evidence, it is expedient to accept the simple mechanism. This is an example of a consecutive reaction, which occurs in two steps, with the intermediate playing a role.

Another piece of evidence for a composite mechanism is the detection of reaction intermediates. In such a case, a reaction scheme must be devised that will account for these intermediates. Sometimes an intermediate can be a fairly stable substance. In other cases the intermediates are unstable species such as atoms and free radicals (fragments of molecules) that subsequently undergo rapid reactions. Free radicals can be detected by spectroscopy and other means. When organic molecules are raised to high temperatures, they decompose into smaller molecules, and organic free radicals have often been detected as intermediates. In an explosion, such as that between hydrogen and oxygen, free radicals such as hydroxyl can be detected.

Composite reaction mechanisms are of various kinds. Aside from the simple consecutive schemes, there are some special mechanisms that give rise to oscillatory behaviour: the amount of a product continuously rises and falls over a period of time. The conditions for this behaviour are that there must be at least two species involved in the reaction and there must be feedback, which means that products of the reaction affect the rate. There are also reaction mechanisms that give rise to what is technically known as chaos, or catastrophe. With such reactions it is impossible to predict the outcome. Chaotic conditions also require that there be feedback and that at least three species be involved.

Sometimes a complex reaction mechanism involves a cycle of reactions such that certain intermediates consumed in one step are regenerated in another. For example, the accepted mechanism of the reaction between hydrogen and bromine, which can be written as:

$H_2 + Br_2 \rightarrow 2HBr$, includes the steps $Br + H_2 \rightarrow HBr + H$ $H + Br_2 \rightarrow HBr + Br$.

In the first of these steps a bromine atom is consumed, but in the second a bromine atom is regenerated. This pair of reactions can thus occur with the production of two molecules of hydrogen bromide, the product of the reaction, without loss of bromine atoms. This pair of reactions is called a cycle of reactions, and it can occur a number of times, in which case the reaction is referred to as a chain reaction. The two reactions in which bromine is regenerated are known as the chain-propagating steps. The average number of times the pair of steps is repeated is known as the chain length.

One necessary condition for a proposed reaction mechanism to be correct is that it must account for the overall kinetic behaviour of the reaction—in particular, for the dependence of the reaction rate on the reactant concentrations. For any proposed reaction mechanism, it is possible to write down equations for the rate of each step in terms of the reactant concentration and then to solve the equations for the overall rate. A practical difficulty arises, since no exact mathematical solution is possible for all except the simplest of mechanisms. If one has values for the rate constants, solutions can be obtained with a computer, but explicit rate equations provide more insight into the reactions. One therefore looks for approximate solutions of the equations. One of these is provided by the steady-state treatment, which is applicable if (and only if) the intermediates are species that can be present only at low concentrations. If this condition is satisfied by an intermediate, the rate of change of its concentration during the course of reaction is always small and, as a good approximation, can be assumed to be zero, which means that the intermediate exists in a steady state. This approximation may safely be applied to atoms and free radicals present as reaction intermediates. With this approximation it is usually possible to obtain a reliable approximate equation for the overall reaction rate in terms of reactant concentrations. If this agrees with the experimental behaviour, the mechanism is accepted.

One situation to which the steady-state treatment does not apply is when a reaction is an explosion. Explosions occur because the concentration of intermediates does not remain steady during the course of reaction but rises to a high value so that the reaction goes out of control. This occurs if the reaction mechanism involves a special kind of chain called a branching chain. In the hydrogen-oxygen explosion, for example, the following reaction is known to occur:

$$H + O_2 \rightarrow OH + O.$$

In this step a single chain carrier hydrogen atom has produced two chain carriers: a hydroxyl group and an oxygen atom. The number of chain carriers increases rapidly and leads to an explosion.

Theories of Reaction Rates

Two different theoretical approaches to chemical kinetics have led to an understanding of the details of how elementary chemical reactions occur. Both of these are based on the idea of potential-energysurfaces, which are models showing how the potential energy of a reaction system varies with certain critical interatomic distances. The course of an elementary reaction is represented by the movement of the system over the potential-energy surface. One theoretical approach to the problem involves studying the region of the potential-energy surface that corresponds to the highest point on the energy barrier that separates the reactants from the products. This approach is relatively simple and leads to explicit general expressions for the reaction rate. The second approach involves considering the dynamics of the motion of the system over the potential-energy surface.

Transition-state Theory

The idea of a potential-energy surface sprang from the ideas of Dutch physical chemist Jacobus Henricus van 't Hoff and Swedish physicist Svante August Arrhenius that were put forward to explain the effect of temperature on reaction rates. An important advance was made in 1931 by

American chemist Henry Eyring and British chemist Michael Polanyi, who constructed, on the basis of quantum mechanics, a potential-energy surface for the simple reaction,

$$H^\alpha + H^\beta - H^\gamma \rightarrow H^\alpha - H^\beta - H^\gamma \rightarrow H^\alpha - H^\beta + H^\gamma.$$

For convenience the labels α, β, and γ are added as superscripts. When this reaction occurs, an atom H^α attacks a hydrogen molecule $H^\beta - H^\gamma$ and abstracts one of the hydrogen atoms from it. As the bond begins to form, the $H^\beta - H^\gamma$ bond becomes more and more extended and finally breaks. Somewhere along the reaction path, there is a particular intermediate state corresponding to the maximum value of the potential energy.

This particular intermediate state is usually designated by the superscript $\ddagger$ (used above in the discussion of the Arrhenius equation). It is known as an activated complex and plays an important role in what has come to be called transition-state theory, developed independently in 1935 by Eyring, Polanyi, and English physical chemist M.G. Evans. The essential feature of the theory is that the activated complexes are considered to be formed from the reactants in a state in which they are in equilibrium with the reactants. Thus, the above reaction can be written as,

$$H^\alpha + H^\beta - H^\gamma \rightarrow H^\alpha - H^\beta - H^{\gamma\ddagger} \text{ (at equilibrium with } H^\alpha + H^\beta - H^\gamma) \rightarrow H^\alpha - H^\beta + H^\gamma.$$

Since the activated complexes are in equilibrium, their concentration can be expressed in terms of the concentrations of the reactants. The reaction rate is this concentration multiplied by the frequency with which they form products, which is known from kinetic theory. Despite the approximations involved in transition-state theory, it has been successful in providing an insight into how chemical reactions occur and how their rates depend on various factors.

Molecular Dynamics

The second theoretical approach to chemical kinetics is referred to as molecular dynamics, or reaction dynamics. It is a more detailed treatment of reactions and is designed to investigate the atomic motions that occur during a chemical reaction and the quantum states of the reactant and product molecules. Such studies are important in testing the validity of transition-state theory and similar treatments. Also, there are important practical applications of kinetics, such as reactions occurring in lasers, for which information about the energy states of the products of a chemical reaction is needed; this information is not provided by transition-state theory but is an important outcome of molecular dynamics.

Consider a simple reaction of the type $A + B - C \rightarrow Y + Z$, where A is an atom and B−C is a diatomic molecule. A dynamical calculation would first involve calculating, using quantum mechanics, a potential-energy surface that gives the potential energy corresponding to a set of initial configurations. One can make dynamical calculations for a variety of vibrational states of the reactant B−C and for a variety of translational energies. Ideally the calculations would be based on quantum mechanics, but this proves difficult, and often classical mechanics is used. Except for certain types of reactions where quantum effects are important, it appears that for many reaction systems the error involved in neglecting quantum effects is not great. In principle the dynamical calculations should give more-reliable results than any other treatment; however, for any but the simplest reactions, the computer calculations are time-consuming, and, since many approximations must be made to save computer time, there is often some uncertainty about the results.

Much work has been done along these lines, and on the whole the agreement with both experiment and transition-state theory is satisfactory. In addition, some calculations suggest important generalizations about reactions. For example, the form of the potential-energy surface greatly influences whether energy released in a reaction resides in the vibrations of the product molecules or in their kinetic energy of translation. Experimental studies of simple reactions, particularly some that involve the study of light emission in reactions (chemiluminescence) and that use narrow molecular beams, have also contributed to knowledge of chemical reactions.

Radiochemistry

Radiochemistry is the chemistry of radioactive materials, where radioactive isotopes of elements are used to study the properties and chemical reactions of non-radioactive isotopes (often within radiochemistry the absence of radioactivity leads to a substance being described as being *inactive* as the isotopes are *stable*). Much of radiochemistry deals with the use of radioactivity to study ordinary chemical reactions. This is very different from radiation chemistry where the radiation levels are kept too low to influence the chemistry.

Radiochemistry includes the study of both natural and man-made radioisotopes.

Main Decay Modes

All radioisotopes are unstable isotopes of elements—undergo nuclear decay and emit some form of radiation. The radiation emitted can be of several types including alpha, beta, gamma radiation, proton and neutron emission along with neutrino and antiparticle emission decay pathways.

- α (alpha) radiation—the emission of an alpha particle (which contains 2 protons and 2 neutrons) from an atomic nucleus. When this occurs, the atom's atomic mass will decrease by 4 units and atomic number will decrease by 2.

- β (beta) radiation—the transmutation of a neutron into an electron and a proton. After this happens, the electron is emitted from the nucleus into the electron cloud.

- γ (gamma) radiation—the emission of electromagnetic energy (such as gamma rays) from the nucleus of an atom. This usually occurs during alpha or beta radioactive decay.

These three types of radiation can be distinguished by their difference in penetrating power.

Alpha can be stopped quite easily by a few centimetres in air or a piece of paper and is equivalent to a helium nucleus. Beta can be cut off by an aluminium sheet just a few millimetres thick and are electrons. Gamma is the most penetrating of the three and is a massless chargeless high energy photon. Gamma radiation requires an appreciable amount of heavy metal radiation shielding (usually lead or barium-based) to reduce its intensity.

Activation Analysis

By neutron irradiation of objects it is possible to induce radioactivity; this activation of stable isotopes to create radioisotopes is the basis of neutron activation analysis. One of the most interesting objects which has been studied in this way is the hair of Napoleon's head, which have been examined for their arsenic content.

A series of different experimental methods exist, these have been designed to enable the measurement of a range of different elements in different matrices. To reduce the effect of the matrix it is common to use the chemical extraction of the wanted element *and/or* to allow the radioactivity due to the matrix elements to decay before the measurement of the radioactivity. Since the matrix effect can be corrected for by observing the decay spectrum, little or no sample preparation is required for some samples, making neutron activation analysis less susceptible to contamination.

The effects of a series of different cooling times can be seen if a hypothetical sample which contains sodium, uranium and cobalt in a 100:10:1 ratio was subjected to a very short pulse of thermal neutrons. The initial radioactivity would be dominated by the ^{24}Na activity (half-life 15 h) but with increasing time the ^{239}Np (half-life 2.4 d after formation from parent ^{239}U with half-life 24 min) and finally the ^{60}Co activity (5.3 yr) would predominate.

Biology Applications

One biological application is the study of DNA using radioactive phosphorus-32. In these experiments stable phosphorus is replaced by the chemical identical radioactive P-32, and the resulting radioactivity is used in analysis of the molecules and their behaviour.

Another example is the work which was done on the methylation of elements such as sulfur, selenium, tellurium and polonium by living organisms. It has been shown that bacteria can convert these elements into volatile compounds, it is thought that methylcobalamin (vitamin B_{12}) alkylates these elements to create the dimethyls. It has been shown that a combination of Cobaloxime and inorganic polonium in sterile water forms a volatile polonium compound, while a control experiment which did not contain the cobalt compound did not form the volatile polonium compound. For the sulfur work the isotope ^{35}S was used, while for polonium ^{207}Po was used. In some related work by the addition of ^{57}Co to the bacterial culture, followed by isolation of the cobalamin from the bacteria (and the measurement of the radioactivity of the isolated cobalamin) it was shown that the bacteria convert available cobalt into methylcobalamin.

In medicine PET (Positron Emission Tomography) scans are commonly used in diagnostic purposes in. A radiative tracer is injected intravenously into the patient and then taken to the PET machine. The radioactive tracer releases radiation outward from the patient and the cameras in the machine interpret the radiation rays from the tracer. PET scan machines use solid state scintillation detection because of its high detection efficiency, NaI(Tl) crystals absorb the tracers radiation and produce photons that get converted into an electrical signal for the machine to analyze.

Environmental

Radiochemistry also includes the study of the behaviour of radioisotopes in the environment; for instance, a forest or grass fire can make radioisotopes become mobile again. In these experiments, fires were started in the exclusion zone around Chernobyl and the radioactivity in the air down-wind was measured.

It is important to note that a vast number of processes are able to release radioactivity into the environment, for example the action of cosmic rays on the air is responsible for the formation of radioisotopes (such as ^{14}C and ^{32}P), the decay of ^{226}Ra forms ^{222}Rn which is a gas which can diffuse through rocks before entering buildings and dissolve in water and thus enter drinking water in addition human activities such as bomb tests, accidents, and normal releases from industry have resulted in the release of radioactivity.

Chemical form of the Actinides

The environmental chemistry of some radioactive elements such as plutonium is complicated by the fact that solutions of this element can undergo disproportionation and as a result many different oxidation states can coexist at once. Some work has been done on the identification of the oxidation state and coordination number of plutonium and the other actinides under different conditions. This includes work on both solutions of relatively simple complexes and work on colloids Two of the key matrixes are soil/rocks and concrete, in these systems the chemical properties of plutonium have been studied using methods such as EXAFS and XANES.

Movement of Colloids

While binding of a metal to the surfaces of the soil particles can prevent its movement through a layer of soil, it is possible for the particles of soil which bear the radioactive metal can migrate as colloidal particles through soil. This has been shown to occur using soil particles labeled with ^{134}Cs, these have been shown to be able to move through cracks in the soil.

Normal Background

Radioactivity is present everywhere (and has been since the formation of the earth). According to the International Atomic Energy Agency, one kilogram of soil typically contains the following amounts of the following three natural radioisotopes 370 Bq ^{40}K (typical range 100–700 Bq), 25 Bq ^{226}Ra (typical range 10–50 Bq), 25 Bq ^{238}U (typical range 10–50 Bq) and 25 Bq ^{232}Th (typical range 7–50 Bq).

Action of Microorganisms

The action of micro-organisms can fix uranium; Thermoanaerobacter can use chromium(VI), iron(III), cobalt(III), manganese(IV) and uranium(VI) as electron acceptors while acetate, glucose, hydrogen, lactate, pyruvate, succinate, and xylose can act as electron donors for the metabolism of the bacteria. In this way the metals can be reduced to form magnetite (Fe_3O_4), siderite ($FeCO_3$), rhodochrosite ($MnCO_3$), and uraninite (UO_2). Other researchers have also worked on the fixing of uranium using bacteria, Francis R. Livens *et al.* (Working at Manchester) have suggested that the reason why *Geobacter sulfurreducens* can reduce UO_2^{2+} cations to uranium dioxide is that the bacteria reduce the uranyl cations to UO_2^+ which then undergoes disproportionation to form UO_2^{2+} and UO_2. This reasoning was based (at least in part) on the observation that NpO_2^+ is not converted to an insoluble neptunium oxide by the bacteria.

Electrochemistry

Electrochemistry is the study of chemical processes that cause electrons to move. This movement of electrons is called electricity, which can be generated by movements of electrons from one element to another in a reaction known as an oxidation-reduction ("redox") reaction. A redox reaction is a reaction that involves a change in oxidation state of one or more elements. When a substance loses an electron, its oxidation state increases; thus, it is oxidized. When a substance gains an electron, its oxidation state decreases, thus being reduced. For example, for the redox reaction,

$$H_2 + F_2 \rightarrow 2\,HF$$

can be rewritten as follows:

- Oxidation reaction:

$$H_2 \rightarrow 2\,H^+ + 2\,e^-$$

- Reduction reaction:

$$F_2 + 2\,e^- \rightarrow 2\,F^-$$

- Overall Reaction:

$$H_2 + F_2 \rightarrow 2\,H^+ + 2\,F^-$$

Oxidation is the loss of electrons, whereas reduction refers to the acquisition of electrons, as illustrated in the respective reactions above. The species being oxidized is also known as the reducing agent or reductant, and the species being reduced is called the oxidizing agent or oxidant. In this case, H_2 is being oxidized (and is the reducing agent), while F_2 is being reduced (and is the oxidizing agent).

Applications of Electrochemistry

Electrochemistry has a number of different uses, particularly in industry. The principles of cells are used to make electrical batteries. In science and technology, a battery is a device that stores chemical energy and makes it available in an electrical form. Batteries are made of electrochemical devices such as one or more galvanic cells or fuel cells. Batteries have many uses including in:

- Torches.
- Electrical appliances such as cellphones (long-life alkaline batteries).
- Digital cameras (lithium batteries).
- Hearing aids (silver-oxide batteries).
- Digital watches (mercury/silver-oxide batteries).
- Military applications (thermal batteries).

We are going to look at a few examples of the uses of electrochemistry in industry.

Electroplating

The electrolytic cell can be used for electroplating.

Electroplating

The process of coating an electrically conductive object with a thin layer of metal using an electrical current.

Electroplating occurs when an electrically conductive object is coated with a layer of metal using electrical current. Sometimes, electroplating is used to give a metal particular properties or for aesthetic reasons:

- Corrosion protection.

- Abrasion and wear resistance.

- The production of jewellery.

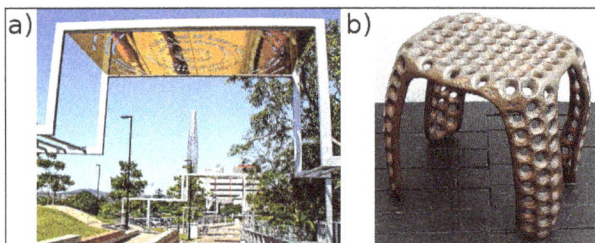

a) An electroplated piece of aluminium artwork and b) a wax stool electroplated in copper.

Electro-refining (also sometimes called electrowinning) is electroplating on a large scale. Copper plays a major role in the electrical industry as it is very conductive and is used in electric cables. One of the problems though is that copper must be pure if it is to be an effective current carrier. One of the methods used to purify copper, is electrowinning (copper ore is processed into impure blister copper, which is then deposited as pure copper through electroplating). The copper electro-winning process is as follows:

- A bar of impure copper containing other metallic impurities acts as the anode.

- The cathode is made up of pure copper with few impurities.

- The electrolyte is a solution of aqueous $CuSO_4$ and H_2SO_4.

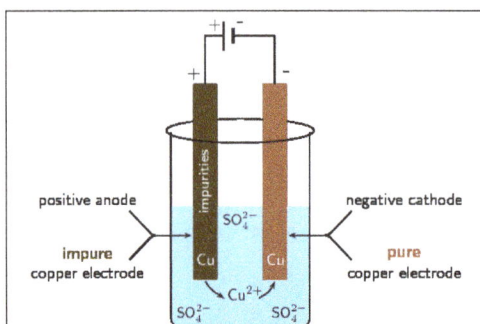

A simplified diagram to illustrate what happens during the electrowinning of copper.

- When current passes through the cell, electrolysis takes place:

  ○ The impure copper anode oxidises to form Cu^{2+} ions in solution. The anode decreases in mass. $Cu(s) \rightarrow Cu^{2+}(aq) + 2e^-$.

  ○ At the cathode reduction of positive copper ions takes place to produce pure copper metal. The cathode increases in mass. $Cu^{2+}(aq) + 2e^- \rightarrow Cu(s)$ (>99% purity).

- The other metal impurities do not dissolve (Au(s), Ag(s)) and form a solid sludge at the bottom of the tank or remain in solution (Zn(aq), Fe(aq) and Pb(aq)) in the electrolyte.

The Chloralkali Industry

The chlorine-alkali (chloralkali) industry is an important part of the chemical industry, which produces chlorine and sodium hydroxide through the electrolysis of the raw material brine. Brine is a saturated solution of sodium chloride (NaCl) that is obtained from natural salt deposits.

Brine

A saturated aqueous solution of sodium chloride.

The products of the chloralkali industry have a number of important uses:

Chlorine is used:

- To purify water.

- As a disinfectant.

- In the production of:

  ○ Hypochlorous acid (used to kill bacteria in drinking water).

  ○ Paper, food.

  ○ Antiseptics, insecticides, medicines, textiles, laboratory chemicals.

  ○ Paints, petroleum products, solvents, plastics (such as polyvinyl chloride).

Sodium hydroxide (also known as 'caustic soda') is used to:

- Make soap and other cleaning agents.

- Purify bauxite (the ore of aluminium).

- Make paper.

- Make rayon (artificial silk).

One of the problems of producing chlorine and sodium hydroxide is that when they are produced together the chlorine combines with the sodium hydroxide to form chlorate (ClO^-) and chloride (Cl^-) ions. This leads to the production of sodium chlorate, NaClO, a component of household bleach.

To overcome this problem the chlorine and sodium hydroxide must be separated from each other so that they don't react. There are three industrial processes that have been designed to overcome this problem. All three methods involve electrolytic cells.

The Mercury Cell In the mercury-cell:

- The anode is a carbon electrode suspended from the top of a chamber.

- The cathode is liquid mercury that flows along the floor of this chamber.

- The electrolyte is brine (NaCl solution) that is passed through the chamber.

- When an electric current is applied to the circuit, chloride ions in the electrolyte are oxidised at the anode to form chlorine gas.

$$2\,Cl^-(aq) \rightarrow Cl_2(g) + 2\,e^-$$

- At the same time sodium ions are reduced at the anode to solid sodium. The solid sodium dissolves in the mercury making a sodium/mercury amalgam.

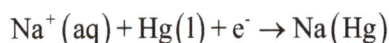

$$Na^+(aq) + Hg(l) + e^- \rightarrow Na(Hg)$$

- The amalgam is poured into a separate vessel, where it decomposes into sodium and mercury.

- The sodium reacts with water in the vessel and produces sodium hydroxide and hydrogen gas, while the mercury returns to the electrolytic cell to be used again.

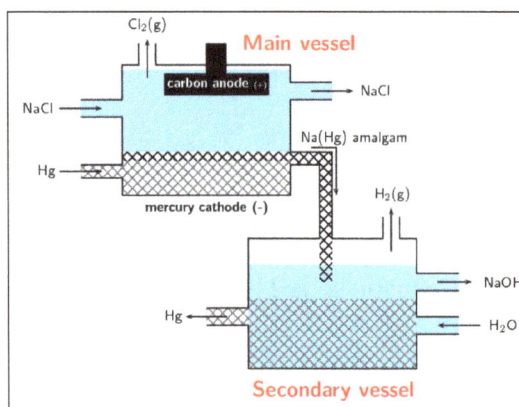

$$2\,Na(Hg) + 2\,H_2O(l) \rightarrow 2\,NaOH(aq) + H_2(g) + Hg(l)$$

The mercury cell.

The following animation gives a good demonstration of how a mercury cell works.

This method only produces a fraction of the chlorine and sodium hydroxide that is used by industry as it has certain disadvantages:

- Mercury is expensive and toxic.

- Some mercury always escapes with the brine that has been used.

- Mercury reacts with the brine to form mercury(II) chloride.

- The mercury cell requires a lot of electricity.

- Although the chlorine gas produced is very pure, mercury has to be removed from the sodium hydroxide and hydrogen gas mixture.

In the past the effluent was released into lakes and rivers, causing mercury to accumulate in fish and other animals feeding on the fish. Today, the brine is treated before it is discharged so that the environmental impact is lower.

The Diaphragm Cell In the diaphragm-cell:

- A porous diaphragm divides the electrolytic cell into an anode compartment and a cathode compartment.

- Brine is introduced into the anode compartment and flows through the diaphragm into the cathode compartment.

- An electric current is passed through the brine causing the salt's chlorine ions and sodium ions to move to the electrodes.

- Chlorine gas is produced at the anode 2Cl–(aq) + 2e–→Cl$_2$(g).

- At the cathode, sodium ions react with water forming caustic soda (NaOH) and hydrogen gas.

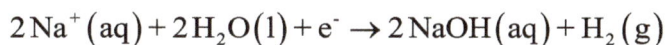

$$2\,Na^+\left(aq\right)+2\,H_2O\left(1\right)+e^- \rightarrow 2\,NaOH\left(aq\right)+H_2\left(g\right)$$

- Some NaCl salt remains in the solution with the caustic soda and can be removed at a later stage.

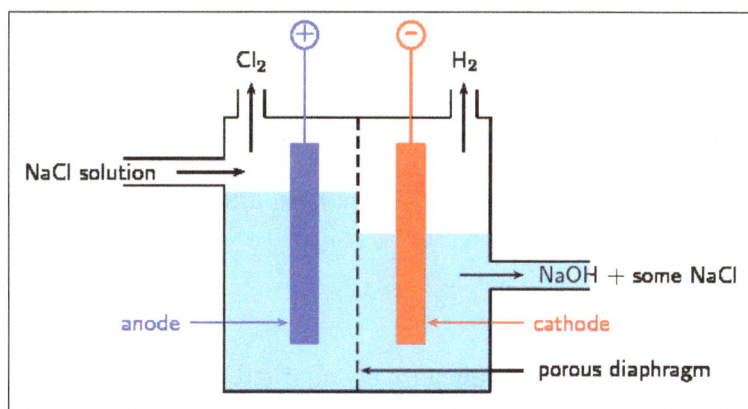

The diaphragm cell.

The following animation gives a good demonstration of how a diaphragm cell works.

The advantages of the diaphragm cell are:

- Uses less energy than the mercury cell.

- Does not contain toxic mercury.

It also has disadvantages however:

- The sodium hydroxide is much less concentrated and not as pure.

- The chlorine gas often contains oxygen gas as well.

- The process is less cost-effective as the sodium hydroxide solution needs to be concentrated and purified before it can be used.

The Membrane Cell

The membrane cell is very similar to the diaphragm cell, and the same reactions occur. The main differences are:

- The two electrodes are separated by an ion-selective membrane, rather than by a diaphragm.

- The membrane structure allows cations to pass through it between compartments of the cell but does not allow anions to pass through (this has nothing to do with the size of the pores, but rather with the charge on the ions).

- Brine is pumped into the anode compartment, and only the positively chargedpositively charged sodium ionssodium ions pass into the cathode compartment, which contains pure water.

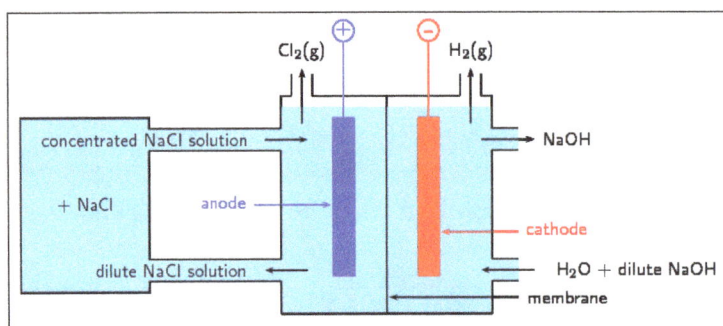

The membrane cell.

The following animation gives a good demonstration of how a membrane cell works.

- At the positively charged anodepositively charged anode, Cl⁻ ions from the brine are oxidised to Cl_2 gas.

$$2Cl^- (aq) \rightarrow Cl_2 (g) + 2e^-$$

- At the negatively charged cathodenegatively charged cathode, hydrogen ions in the water are reduced to hydrogen gas.

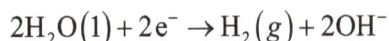

$$2H_2O(l) + 2e^- \rightarrow H_2(g) + 2OH^-$$

- The Na+Na+ ions flow through the membrane to the cathode compartment and react with the remaining hydroxide (OH⁻) ions from the water to form sodium hydroxide (NaOH).

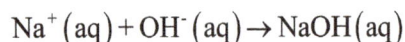

$$Na^+ (aq) + OH^- (aq) \rightarrow NaOH(aq)$$

- The chloride ions cannot pass through the membrane, so the chlorine does not come into contact with the sodium hydroxide in the cathode compartment. The sodium hydroxide is removed from the cell. The overall equation is as follows:

$$2NaCl(aq) + 2H_2O(l) \rightarrow Cl_2(g) + H_2(g) + 2NaOH(g)$$

The advantages of using this method are:

- The sodium hydroxide that is produced is very pure because it is kept separate from the sodium chloride solution.

- The sodium hydroxide has a relatively high concentration.

- This process uses the least electricity of all three cells.

- The cell is cheaper to operate than the other two cells.

- The cell does not contain toxic mercury or asbestos.

The Extraction of Aluminium

Aluminium is a commonly used metal in industry, where its properties of being both light and strong can be utilised. It is used in the manufacture of products such as aeroplanes and motor cars. The metal is present in deposits of bauxite. Bauxite is a mixture of silicas, iron oxides and hydrated alumina $(Al_2O_3.\times H_2O)$.

Electrolysis can be used to extract aluminium from bauxite. The process described below produces 99% pure aluminium:

- Aluminium is melted along with cryolite (Na_3AlF_6) which acts as the electrolyte. Cryolite helps to lower the melting point and dissolve the ore.

- The carbon rod anode provides a site for the oxidation of O^{2-} and F^- ions. Oxygen and fluorine gas are given off at the anode and also result in anode consumption.

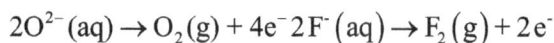

$$2O^{2-}(aq) \rightarrow O_2(g) + 4e^- \quad 2F^-(aq) \rightarrow F_2(g) + 2e^-$$

- At the cathode cell lining, the Al^{3+} ions are reduced and metal aluminium deposits on the lining. $Al^{3+}(aq) + 3e^- \rightarrow Al(s)$ (99% purity).

- The AlF_6^{3-} electrolyte is stable and remains in its molten state.

The overall reaction is as follows:

$$2Al_2O_3(s) \rightarrow 4Al(s) + 3O_2(g)$$

The only problem with this process is that the reaction is endothermic and large amounts of electricity are needed to drive the reaction. The process is therefore very expensive.

Thermochemistry

Thermochemistry is the study of the heat energy associated with chemical reactions and/or physical transformations. A reaction may release or absorb energy, and a phase change may do the same, such as in melting and boiling. Thermochemistry focuses on these energy changes, particularly on the system's energy exchange with its surroundings. Thermochemistry is useful in predicting reactant and product quantities throughout the course of a given reaction. In combination with entropy determinations, it is also used to predict whether a reaction is spontaneous or non-spontaneous, favorable or unfavorable.

Endothermic reactions absorb heat, while exothermic reactions release heat. Thermochemistry coalesces the concepts of thermodynamics with the concept of energy in the form of chemical bonds. The subject commonly includes calculations of such quantities as heat capacity, heat of combustion, heat of formation, enthalpy, entropy, free energy, and calories.

Calorimetry

The measurement of heat changes is performed using calorimetry, usually an enclosed chamber within which the change to be examined occurs. The temperature of the chamber is monitored either using a thermometer or thermocouple, and the temperature plotted against time to give a graph from which fundamental quantities can be calculated. Modern calorimeters are frequently supplied with automatic devices to provide a quick read-out of information, one example being the differential scanning calorimeter (DSC).

Systems

Several thermodynamic definitions are very useful in thermochemistry. A system is the specific portion of the universe that is being studied. Everything outside the system is considered the surroundings or environment. A system may be:

- A (completely) isolated system which can exchange neither energy nor matter with the surroundings, such as an insulated bomb calorimeter.

- A thermally isolated system which can exchange mechanical work but not heat or matter, such as an insulated closed piston or balloon.

- A mechanically isolated system which can exchange heat but not mechanical work or matter, such as an uninsulated bomb calorimeter.

- A closed system which can exchange energy but not matter, such as an uninsulated closed piston or balloon.

- An open system which it can exchange both matter and energy with the surroundings, such as a pot of boiling water.

Processes

A system undergoes a process when one or more of its properties changes. A process relates to the change of state. An isothermal (same-temperature) process occurs when temperature of the system remains constant. An isobaric (same-pressure) process occurs when the pressure of the system remains constant. A process is adiabatic when no heat exchange occurs.

Environmental Chemistry

Environmental chemistry is the scientific study of the chemical and biochemical phenomena that occur in natural places. It should not be confused with green chemistry, which seeks to reduce potential pollution at its source. It can be defined as the study of the sources, reactions, transport, effects, and fates of chemical species in the air, soil, and water environments; and the effect of human activity and biological activity on these. Environmental chemistry is an interdisciplinary science that includes atmospheric, aquatic and soil chemistry, as well as heavily relying on analytical chemistry and being related to environmental and other areas of science.

White bags filled with contaminated stones line the shore near an industrial oil spill.

Environmental chemistry is the study of chemical processes occurring in the environment which are impacted by humankind's activities. These impacts may be felt on a local scale, through the presence of urban cities' air pollutants or toxic substances arising from a chemical waste site, or on a global scale, through depletion of stratospheric ozone or global warming. The focus in our courses and research activities is upon developing a fundamental understanding of the nature of these chemical processes, so that humankind's activities can be accurately evaluated.

Environmental chemistry involves first understanding how the uncontaminated environment works, which chemicals in what concentrations are present naturally, and with what effects. Without this it would be impossible to accurately study the effects humans have on the environment through the release of chemicals.

Environmental chemists draw on a range of concepts from chemistry and various environmental sciences to assist in their study of what is happening to a chemical species in the environment. Important general concepts from chemistry include understanding chemical reactions and equations, solutions, units, sampling, and analytical techniques.

Contamination

A contaminant is a substance present in nature at a level higher than fixed levels or that would not otherwise be there. This may be due to human activity and bioactivity. The term contaminant is often used interchangeably with *pollutant*, which is a substance that has a detrimental impact on the surrounding environment. Whilst a contaminant is sometimes defined as a substance present in the environment as a result of human activity, but without harmful effects, it is sometimes the case that toxic or harmful effects from contamination only become apparent at a later date.

The "medium" (e.g. soil) or organism (e.g. fish) affected by the pollutant or contaminant is called a *receptor*, whilst a *sink* is a chemical medium or species that retains and interacts with the pollutant e.g. as carbon sink and its effects by microbes.

Environmental Indicators

Chemical measures of water quality include dissolved oxygen (DO), chemical oxygen demand (COD), biochemical oxygen demand (BOD), total dissolved solids (TDS), pH, nutrients (nitrates and phosphorus), heavy metals, soil chemicals (including copper, zinc, cadmium, lead and mercury), and pesticides.

Applications

Environmental chemistry is used by the Environment Agency (in England and Wales), the United States Environmental Protection Agency, the Association of Public Analysts, and other environmental agencies and research bodies around the world to detect and identify the nature and source of pollutants. These can include:

- Heavy metal contamination of land by industry. These can then be transported into water bodies and be taken up by living organisms.

- Nutrients leaching from agricultural land into water courses, which can lead to algal blooms and eutrophication.

- Urban runoff of pollutants washing off impervious surfaces (roads, parking lots, and rooftops) during rain storms. Typical pollutants include gasoline, motor oil and other hydrocarbon compounds, metals, nutrients and sediment (soil).

- Organometallic compounds.

Methods

Quantitative chemical analysis is a key part of environmental chemistry, since it provides the data that frame most environmental studies.

Common analytical techniques used for quantitative determinations in environmental chemistry include classical wet chemistry, such as gravimetric, titrimetric and electrochemical methods. More sophisticated approaches are used in the determination of trace metals and organic compounds. Metals are commonly measured by atomic spectroscopy and mass spectrometry: Atomic Absorption Spectrophotometry (AAS) and Inductively Coupled Plasma Atomic Emission (ICP-AES) or Inductively Coupled Plasma Mass Spectrometric (ICP-MS) techniques. Organic compounds are commonly measured also using mass spectrometric methods, such as Gas chromatography-mass spectrometry (GC/MS) and Liquid chromatography-mass spectrometry (LC/MS). Tandem Mass spectrometry MS/MS and High Resolution/Accurate Mass spectrometry HR/AM offer sub part per trillion detection. Non-MS methods using GCs and LCs having universal or specific detectors are still staples in the arsenal of available analytical tools.

Other parameters often measured in environmental chemistry are radiochemicals. These are pollutants which emit radioactive materials, such as alpha and beta particles, posing danger to human health and the environment. Particle counters and Scintillation counters are most commonly used for these measurements. Bioassays and immunoassays are utilized for toxicity evaluations of chemical effects on various organisms. Polymerase Chain Reaction PCR is able to identify species of bacteria and other organisms through specific DNA and RNA gene isolation and amplification and is showing promise as a valuable technique for identifying environmental microbial contamination.

Nuclear Chemistry

Nuclear chemistry is the subfield of chemistry dealing with radioactivity, nuclear processes, and transformations in the nuclei of atoms, such as nuclear transmutation, and nuclear properties.

It is the chemistry of radioactive elements such as the actinides, radium and radon together with the chemistry associated with equipment (such as nuclear reactors) which are designed to perform nuclear processes. This includes the corrosion of surfaces and the behavior under conditions of both normal and abnormal operation (such as during an accident). An important area is the behavior of objects and materials after being placed into a nuclear waste storage or disposal site.

It includes the study of the chemical effects resulting from the absorption of radiation within living animals, plants, and other materials. The radiation chemistry controls much of radiation biology as radiation has an effect on living things at the molecular scale, to explain it another way the radiation alters the biochemicals within an organism, the alteration of the biomolecules then changes the chemistry which occurs within the organism, this change in chemistry then can lead to a biological outcome. As a result, nuclear chemistry greatly assists the understanding of medical treatments (such as cancer radiotherapy) and has enabled these treatments to improve.

It includes the study of the production and use of radioactive sources for a range of processes. These include radiotherapy in medical applications; the use of radioactive tracers within industry, science and the environment; and the use of radiation to modify materials such as polymers.

It also includes the study and use of nuclear processes in *non-radioactive* areas of human activity. For instance, nuclear magnetic resonance (NMR) spectroscopy is commonly used in synthetic organic chemistry and physical chemistry and for structural analysis in macromolecular chemistry.

Nuclear chemistry concerned with the study of nucleus, changes occurring in the nucleus, properties of the particles present in the nucleus and the emission or absorption of radiation from the nucleus.

Main Areas

Radiochemistry is the chemistry of radioactive materials, in which radioactive isotopes of elements are used to study the properties and chemical reactions of non-radioactive isotopes (often within radiochemistry the absence of radioactivity leads to a substance being described as being *inactive* as the isotopes are *stable*).

Radiation Chemistry

Radiation chemistry is the study of the chemical effects of radiation on the matter; this is very different from radiochemistry as no radioactivity needs to be present in the material which is being chemically changed by the radiation. An example is the conversion of water into hydrogen gas and hydrogen peroxide. Prior to radiation chemistry, it was commonly believed that pure water could not be destroyed.

Initial experiments were focused on understanding the effects of radiation on matter. Using a X-ray generator, Hugo Fricke studied the biological effects of radiation as it became a common treatment option and diagnostic method. Fricke proposed and subsequently proved that the energy from X - rays were able to convert water into activated water, allowing it to react with dissolved species.

Chemistry for Nuclear Power

Radiochemistry, radiation chemistry and nuclear chemical engineering play a very important role for uranium and thorium fuel precursors synthesis, starting from ores of these elements, fuel fabrication, coolant chemistry, fuel reprocessing, radioactive waste treatment and storage, monitoring of radioactive elements release during reactor operation and radioactive geological storage, etc.

Study of Nuclear Reactions

A combination of radiochemistry and radiation chemistry is used to study nuclear reactions such as fission and fusion. Some early evidence for nuclear fission was the formation of a short-lived radioisotope of barium which was isolated from neutron irradiated uranium (^{139}Ba, with a half-life of 83 minutes and ^{140}Ba, with a half-life of 12.8 days, are major fission products of uranium). At the time, it was thought that this was a new radium isotope, as it was then standard radiochemical practice to use a barium sulfate carrier precipitate to assist in the isolation of radium.] More recently, a combination of radiochemical methods and nuclear physics has been used to try to make new 'superheavy' elements; it is thought that islands of relative stability exist where the nuclides have half-lives of years, thus enabling weighable amounts of the new elements to be isolated.

The Nuclear Fuel Cycle

This is the chemistry associated with any part of the nuclear fuel cycle, including nuclear reprocessing. The fuel cycle includes all the operations involved in producing fuel, from mining, ore processing and enrichment to fuel production (*Front-end of the cycle*). It also includes the 'in-pile' behavior (use of the fuel in a reactor) before the *back end* of the cycle. The *back end* includes the management of the used nuclear fuel in either a spent fuel pool or dry storage, before it is disposed of into an underground waste store or reprocessed.

Normal and Abnormal Conditions

The nuclear chemistry associated with the nuclear fuel cycle can be divided into two main areas, one area is concerned with operation under the intended conditions while the other area is concerned with maloperation conditions where some alteration from the normal operating conditions has occurred or (*more rarely*) an accident is occurring. Without this process, none of this would be true.

Reprocessing

Law

In the United States, it is normal to use fuel once in a power reactor before placing it in a waste store. The long-term plan is currently to place the used civilian reactor fuel in a deep store. This non-reprocessing policy was started in March 1977 because of concerns about nuclear weapons proliferation. President Jimmy Carter issued a Presidential directive which indefinitely suspended the commercial reprocessing and recycling of plutonium in the United States. This directive was likely an attempt by the United States to lead other countries by example, but many other nations continue to reprocess spent nuclear fuels. The Russian government under President Vladimir Putin repealed a law which had banned the import of used nuclear fuel, which makes it possible for Russians to offer a reprocessing service for clients outside Russia (similar to that offered by BNFL).

PUREX Chemistry

The current method of choice is to use the PUREX liquid-liquid extraction process which uses a tributyl phosphate/hydrocarbon mixture to extract both uranium and plutonium from nitric acid. This extraction is of the nitrate salts and is classed as being of a solvation mechanism. For example, the extraction of plutonium by an extraction agent (S) in a nitrate medium occurs by the following reaction.

$$Pu^{4+}{}_{aq} + 4NO_3{}^-{}_{aq} + 2S_{organic} \rightarrow [Pu(NO_3)_4S_2]_{organic}$$

A complex bond is formed between the metal cation, the nitrates and the tributyl phosphate, and a model compound of a dioxouranium(VI) complex with two nitrates and two triethyl phosphates has been characterised by X-ray crystallography.

When the nitric acid concentration is high the extraction into the organic phase is favored, and when the nitric acid concentration is low the extraction is reversed (the organic phase is *stripped* of the metal). It is normal to dissolve the used fuel in nitric acid, after the removal of the insoluble matter the uranium and plutonium are extracted from the highly active liquor. It is normal to then back extract the loaded organic phase to create a *medium active* liquor which contains mostly uranium and plutonium with only small traces of fission products. This medium active aqueous mixture is then extracted again by tributyl phosphate/hydrocarbon to form a new organic phase, the metal bearing organic phase is then stripped of the metals to form an aqueous mixture of only uranium and plutonium. The two stages of extraction are used to improve the purity of the actinide product, the organic phase used for the first extraction will suffer a far greater dose of radiation. The radiation can degrade the tributyl phosphate into dibutyl hydrogen phosphate. The dibutyl hydrogen phosphate can act as an extraction agent for both the actinides and other metals such as ruthenium. The dibutyl hydrogen phosphate can make the system behave in a more complex manner as it tends to extract metals by an ion exchange mechanism (extraction favoured by low acid concentration), to reduce the effect of the dibutyl hydrogen phosphate it is common for the used organic phase to be washed with sodium carbonate solution to remove the acidic degradation products of the tributyl phosphate.

New Methods being Considered for Future use

The PUREX process can be modified to make a UREX (*UR*anium *EX*traction) process which could be used to save space inside high level nuclear waste disposal sites, such as Yucca Mountain nuclear waste repository, by removing the uranium which makes up the vast majority of the mass and volume of used fuel and recycling it as reprocessed uranium.

The UREX process is a PUREX process which has been modified to prevent the plutonium being extracted. This can be done by adding a plutonium reductant before the first metal extraction step. In the UREX process, ~99.9% of the uranium and >95% of technetium are separated from each other and the other fission products and actinides. The key is the addition of acetohydroxamic acid (AHA) to the extraction and scrubs sections of the process. The addition of AHA greatly diminishes the extractability of plutonium and neptunium, providing greater proliferation resistance than with the plutonium extraction stage of the PUREX process.

Adding a second extraction agent, octyl(phenyl)-*N*,*N*-dibutyl carbamoylmethyl phosphine oxide (CMPO) in combination with tributylphosphate, (TBP), the PUREX process can be turned into the TRUEX (*TR*ans*U*ranic *EX*traction) process this is a process which was invented in the USA by Argonne National Laboratory, and is designed to remove the transuranic metals (Am/Cm) from waste. The idea is that by lowering the alpha activity of the waste, the majority of the waste can then be disposed of with greater ease. In common with PUREX this process operates by a solvation mechanism.

As an alternative to TRUEX, an extraction process using a malondiamide has been devised. The DIAMEX (*DIAM*ide*EX*traction) process has the advantage of avoiding the formation of organic waste which contains elements other than carbon, hydrogen, nitrogen, and oxygen. Such an organic waste can be burned without the formation of acidic gases which could contribute to acid rain. The DIAMEX process is being worked on in Europe by the French CEA. The process is sufficiently mature that an industrial plant could be constructed with the existing knowledge of the process. In common with PUREX this process operates by a solvation mechanism.

Selective Actinide Extraction (SANEX). As part of the management of minor actinides, it has been proposed that the lanthanides and trivalent minor actinides should be removed from the PUREX raffinate by a process such as DIAMEX or TRUEX. In order to allow the actinides such as americium to be either reused in industrial sources or used as fuel the lanthanides must be removed. The lanthanides have large neutron cross sections and hence they would poison a neutron-driven nuclear reaction. To date, the extraction system for the SANEX process has not been defined, but currently, several different research groups are working towards a process. For instance, the French CEA is working on a bis-triaiznyl pyridine (BTP) based process.

Other systems such as the dithiophosphinic acids are being worked on by some other workers.

This is the *UNiversal EX*traction process which was developed in Russia and the Czech Republic, it is a process designed to remove all of the most troublesome (Sr, Cs and minor actinides) radioisotopes from the raffinates left after the extraction of uranium and plutonium from used nuclear fuel. The chemistry is based upon the interaction of caesium and strontium with poly ethylene oxide (poly ethylene glycol) and a cobalt carborane anion (known as chlorinated cobalt dicarbollide). The actinides are extracted by CMPO, and the diluent is a polar aromatic such as nitrobenzene. Other dilents such as *meta*-nitrobenzotrifluoride and phenyl trifluoromethyl sulfone have been suggested as well.

Absorption of Fission Products on Surfaces

Another important area of nuclear chemistry is the study of how fission products interact with surfaces; this is thought to control the rate of release and migration of fission products both from waste containers under normal conditions and from power reactors under accident conditions. Like chromate and molybdate, the $^{99}TcO_4$ anion can react with steel surfaces to form a corrosion resistant layer. In this way, these metaloxo anions act as anodic corrosion inhibitors. The formation of $^{99}TcO_2$ on steel surfaces is one effect which will retard the release of ^{99}Tc from nuclear waste drums and nuclear equipment which has been lost before decontamination (e.g. submarine reactors lost at sea). This $^{99}TcO_2$ layer renders the steel surface passive, inhibiting the anodic corrosion reaction. The radioactive nature of technetium makes this corrosion protection impractical in almost all situations. It has also been shown that $^{99}TcO_4$ anions react to form a layer on the surface of activated carbon (charcoal) or aluminium. A short review of the biochemical properties of a series of key long lived radioisotopes can be read on line.

^{99}Tc in nuclear waste may exist in chemical forms other than the $^{99}TcO_4$ anion, these other forms have different chemical properties. Similarly, the release of iodine-131 in a serious power reactor accident could be retarded by absorption on metal surfaces within the nuclear plant.

Spinout Areas

Some methods first developed within nuclear chemistry and physics have become so widely used within chemistry and other physical sciences that they may be best thought of as separate from *normal* nuclear chemistry. For example, the isotope effect is used so extensively to investigate chemical mechanisms and the use of cosmogenic isotopes and long-lived unstable isotopes in geology that it is best to consider much of isotopic chemistry as separate from nuclear chemistry.

Kinetics (use within Mechanistic Chemistry)

The mechanisms of chemical reactions can be investigated by observing how the kinetics of a reaction is changed by making an isotopic modification of a substrate, known as the kinetic isotope effect. This is now a standard method in organic chemistry. Briefly, replacing normal hydrogen (protons) by deuterium within a molecule causes the molecular vibrational frequency of X-H (for example C-H, N-H and O-H) bonds to decrease, which leads to a decrease in vibrational zero-point energy. This can lead to a decrease in the reaction rate if the rate-determining step involves breaking a bond between hydrogen and another atom. Thus, if the reaction changes in rate when protons are replaced by deuteriums, it is reasonable to assume that the breaking of the bond to hydrogen is part of the step which determines the rate.

Uses within Geology, Biology and Forensic Science

Cosmogenic isotopes are formed by the interaction of cosmic rays with the nucleus of an atom. These can be used for dating purposes and for use as natural tracers. In addition, by careful measurement of some ratios of stable isotopes it is possible to obtain new insights into the origin of bullets, ages of ice samples, ages of rocks, and the diet of a person can be identified from a hair or other tissue sample.

Biology

Within living things, isotopic labels (both radioactive and nonradioactive) can be used to probe how the complex web of reactions which makes up the metabolism of an organism converts one substance to another. For instance a green plant uses light energy to convert water and carbon dioxide into glucose by photosynthesis. If the oxygen in the water is labeled, then the label appears in the oxygen gas formed by the plant and not in the glucose formed in the chloroplasts within the plant cells.

For biochemical and physiological experiments and medical methods, a number of specific isotopes have important applications.

- Stable isotopes have the advantage of not delivering a radiation dose to the system being studied; however, a significant excess of them in the organ or organism might still interfere with its functionality, and the availability of sufficient amounts for whole-animal studies is limited for many isotopes. Measurement is also difficult, and usually requires mass spectrometry to determine how much of the isotope is present in particular compounds, and there is no means of localizing measurements within the cell.

- ^2H (deuterium), the stable isotope of hydrogen, is a stable tracer, the concentration of which can be measured by mass spectrometry or NMR. It is incorporated into all cellular structures. Specific deuterated compounds can also be produced.

- ^{15}N, the stable isotope of nitrogen, has also been used. It is incorporated mainly into proteins.

- Radioactive isotopes have the advantages of being detectable in very low quantities, in being easily measured by scintillation counting or other radiochemical methods, and in being localizable to particular regions of a cell, and quantifiable by autoradiography. Many compounds with the radioactive atoms in specific positions can be prepared, and are widely available commercially. In high quantities they require precautions to guard the workers from the effects of radiation—and they can easily contaminate laboratory glassware and other equipment. For some isotopes the half-life is so short that preparation and measurement is difficult.

By organic synthesis it is possible to create a complex molecule with a radioactive label that can be confined to a small area of the molecule. For short-lived isotopes such as ^{11}C, very rapid synthetic methods have been developed to permit the rapid addition of the radioactive isotope to the molecule. For instance a palladium catalysed carbonylation reaction in a microfluidic device has been used to rapidly form amides and it might be possible to use this method to form radioactive imaging agents for PET imaging.

- ^{3}H, Tritium, the radioisotope of hydrogen, is available at very high specific activities, and compounds with this isotope in particular positions are easily prepared by standard chemical reactions such as hydrogenation of unsaturated precursors. The isotope emits very soft beta radiation, and can be detected by scintillation counting.

- ^{11}C, Carbon-11 is usually produced by cyclotron bombardment of ^{14}N with protons. The resulting nuclear reaction is ^{14}N(p,α)^{11}C. Additionally, Carbon-11 can also be made using a cyclotron; boron in the form of boric oxide is reacted with protons in a (p,n) reaction. Another alternative route is to react ^{10}B with deuterons. By rapid organic synthesis, the ^{11}C compound formed in the cyclotron is converted into the imaging agent which is then used for PET.

- ^{14}C, Carbon-14 can be made (as above), and it is possible to convert the target material into simple inorganic and organic compounds. In most organic synthesis work it is normal to try to create a product out of two approximately equal sized fragments and to use a convergent route, but when a radioactive label is added, it is normal to try to add the label late in the synthesis in the form of a very small fragment to the molecule to enable the radioactivity to be localised in a single group. Late addition of the label also reduces the number of synthetic stages where radioactive material is used.

- ^{18}F, fluorine-18 can be made by the reaction of neon with deuterons, ^{20}Ne reacts in a (d,^{4}He) reaction. It is normal to use neon gas with a trace of stable fluorine (^{19}F$_2$). The ^{19}F$_2$ acts as a carrier which increases the yield of radioactivity from the cyclotron target by reducing the amount of radioactivity lost by absorption on surfaces. However, this reduction in loss is at the cost of the specific activity of the final product.

Nuclear Magnetic Resonance (NMR)

NMR spectroscopy uses the net spin of nuclei in a substance upon energy absorption to identify molecules. This has now become a standard spectroscopic tool within synthetic chemistry. One major use of NMR is to determine the bond connectivity within an organic molecule.

NMR imaging also uses the net spin of nuclei (commonly protons) for imaging. This is widely used for diagnostic purposes in medicine, and can provide detailed images of the inside of a person without inflicting any radiation upon them. In a medical setting, NMR is often known simply as "magnetic resonance" imaging, as the word 'nuclear' has negative connotations for many people.

Medicinal Chemistry

Medicinal chemistry and pharmaceutical chemistry are disciplines at the intersection of chemistry, especially synthetic organic chemistry, and pharmacology and various other biological specialties, where they are involved with design, chemical synthesis and development for market of pharmaceutical agents, or bio-active molecules (drugs).

Compounds used as medicines are most often organic compounds, which are often divided into the broad classes of small organic molecules (e.g., atorvastatin, fluticasone, clopidogrel) and "biologics" (infliximab, erythropoietin, insulin glargine), the latter of which are most often medicinal preparations of proteins (natural and recombinant antibodies, hormones, etc.). Inorganic and organometallic compounds are also useful as drugs (e.g., lithium and platinum-based agents such as lithium carbonate and cisplatin as well as gallium).

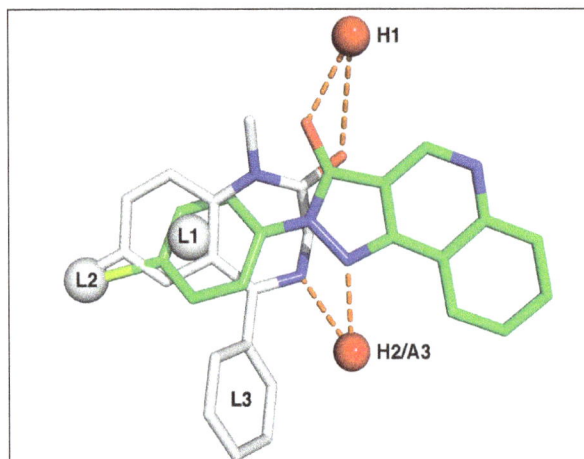

Medicinal chemistry seeks to develop therapeutic agents. Pharmacophore model of the benzodiazepine binding site on the GABAA receptor.

In particular, medicinal chemistry in its most common practice—focusing on small organic molecules—encompasses synthetic organic chemistry and aspects of natural products and computational chemistry in close combination with chemical biology, enzymology and structural biology, together aiming at the discovery and development of new therapeutic agents. Practically speaking, it involves chemical aspects of identification, and then systematic, thorough synthetic alteration of new chemical entities to make them suitable for therapeutic use. It includes synthetic and computational aspects of the study of existing drugs and agents in development in relation to their bioactivities (biological activities and properties), i.e., understanding their structure-activity relationships (SAR). Pharmaceutical chemistry is focused on quality aspects of medicines and aims to assure fitness for purpose of medicinal products.

At the biological interface, medicinal chemistry combines to form a set of highly interdisciplinary sciences, setting its organic, physical, and computational emphases alongside biological areas such as biochemistry, molecular biology, pharmacognosy and pharmacology, toxicology and veterinary and human medicine; these, with project management, statistics, and pharmaceutical business practices, systematically oversee altering identified chemical agents such that after pharmaceutical formulation, they are safe and efficacious, and therefore suitable for use in treatment of disease.

In the Path of Drug Discovery

Discovery is the identification of novel active chemical compounds, often called "hits", which are typically found by assay of compounds for a desired biological activity. Initial hits can come from repurposing existing agents toward a new pathologic processes, and from observations of biologic effects of new or existing natural products from bacteria, fungi, plants, etc. In addition, hits also routinely originate from structural observations of small molecule "fragments" bound to therapeutic targets (enzymes, receptors, etc.), where the fragments serve as starting points to develop more chemically complex forms by synthesis. Finally, hits also regularly originate from en-masse testing of chemical compounds against biological targets, where the compounds may be from novel synthetic chemical libraries known to have particular properties (kinase inhibitory activity, diversity or drug-likeness, etc.), or from historic chemical compound collections or libraries created through combinatorial chemistry. While a number of approaches toward the identification and development of hits exist, the most successful techniques are based on chemical and biological intuition developed in team environments through years of rigorous practice aimed solely at discovering new therapeutic agents.

Hit to Lead and Lead Optimization

Further chemistry and analysis is necessary, first to identify the "triage" compounds that do not provide series displaying suitable SAR and chemical characteristics associated with long-term potential for development, then to improve remaining hit series with regard to the desired primary activity, as well as secondary activities and physiochemical properties such that the agent will be useful when administered in real patients. In this regard, chemical modifications can improve the recognition and binding geometries (pharmacophores) of the candidate compounds, and so their affinities for their targets, as well as improving the physicochemical properties of the molecule that underlie necessary pharmacokinetic/pharmacodynamic (PK/PD), and toxicologic profiles (stability toward metabolic degradation, lack of geno-, hepatic, and cardiac toxicities, etc.) such that the chemical compound or biologic is suitable for introduction into animal and human studies.

Process Chemistry and Development

The final synthetic chemistry stages involve the production of a lead compound in suitable quantity and quality to allow large scale animal testing, and then human clinical trials. This involves the optimization of the synthetic route for bulk industrial production, and discovery of the most suitable drug formulation. The former of these is still the bailiwick of medicinal chemistry, the latter brings in the specialization of formulation science (with its components of physical and polymer chemistry and materials science). The synthetic chemistry specialization in medicinal chemistry aimed at adaptation and optimization of the synthetic route for industrial scale syntheses of hundreds of kilograms or more is termed process synthesis, and involves thorough knowledge of acceptable synthetic practice in the context of large scale reactions (reaction thermodynamics, economics, safety, etc.). Critical at this stage is the transition to more stringent GMP requirements for material sourcing, handling, and chemistry.

Synthetic Analysis

The synthetic methodology employed in medicinal chemistry is subject to constraints that do not apply to traditional organic synthesis. Owing to the prospect of scaling the preparation, safety is of paramount importance. The potential toxicity of reagents affects methodology.

Structural Analysis

The structures of pharmaceuticals are assessed in many ways, in part as a means to predict efficacy, stability, and accessibility. Lipinski's rule of five focus on the number of hydrogen bond donors and acceptors, number of rotatable bonds, surface area, and lipophilicity. Other parameters by which medicinal chemists assess or classify their compounds are: synthetic complexity, chirality, flatness, and aromatic ring count.

References

- Skoog, Douglas A.; Holler, F. James; Crouch, Stanley R. (2007). Principles of Instrumental Analysis. Belmont, CA: Brooks/Cole, Thomson. P. 1. ISBN 978-0-495-01201-6

- Physical-Chemistry, Ny-Pi: chemistryexplained.com, Retrieved 26 August, 2019

- Skoog, Douglas A.; West, Donald M.; Holler, F. James; Crouch, Stanley R. (2014). Fundamentals of Analytical Chemistry. Belmont: Brooks/Cole, Cengage Learning. P. 1. ISBN 978-0-495-55832-3

- Chemical-kinetics, science: britannica.com, Retrieved 10 May, 2019

- Saha, Gopal B. (2010). "PET Scanning Systems". Basics of PET Imaging. Springer, New York, NY. Pp. 19–39. Doi:10.1007/978-1-4419-0805-6_2. ISBN 9781441908049

- Basics-of-Electrochemistry, Electrochemistry, Supplemental-Modules-(Analytical-Chemistry), Analytical-Chemistry: libretexts.org, Retrieved 4 June, 2019

- Vanloon, Gary W.; Duffy, Stephen J. (2000). Environmental Chemistry. Oxford: Oxford. P. 7. ISBN 0-19-856440-6

- 13-electrochemical-reactions-07, electrochemical-reactions, grade-12, science, read: siyavula.com, Retrieved 5 April, 2019

- Andrew Davis, Simon E Ward, ed. (2015). Handbook of Medicinal Chemistry: Principles and Practice Editors. Royal Society of Chemistry. Doi:10.1039/9781782621836. ISBN 978-1-78262-419-6

3
Organic Chemistry

The sub-discipline of chemistry that studies the properties, structures and reactions of compounds that contain carbon in covalent bonding is known as organic chemistry. Some of the major areas of study within this field are functional groups and aliphatic compounds. All the diverse areas of organic chemistry have been carefully analyzed in this chapter.

Organic chemistry is the chemistry of carbon compounds. All organic compounds contain carbon; however, there are some compounds of carbon that are not classified as organic. For example, salts such as carbonates (e.g., $Na_2 CO_3$, $CaCO_3$) and cyanides (e.g., NaCN, KCN) are usually designated as inorganic. Perhaps a more useful description might be: Organic compounds are compounds of carbon that usually contain hydrogen and that may also contain other elements such as oxygen, nitrogen, sulfur, phosphorus, or halogen (F, Cl, Br, or I). In any case, there are very few carbon compounds that are not organic, while there are millions that are.

Scope of Organic Chemistry

The field of organic chemistry includes more than twenty million compounds for which properties have been determined and recorded in the literature. Many hundreds of new compounds are added every day. Much more than half of the world's chemists are organic chemists. Some new organic compounds are simply isolated from plants or animals; some are made by modifying naturally occurring chemicals; but most new organic compounds are actually synthesized in the laboratory from other (usually smaller) organic molecules. Over the years organic chemists have developed a broad array of reactions that allow them to make all kinds of complex products from simpler starting materials.

Singular Attributes of Carbon

When one considers the millions of chemical compounds that are known and notes that more than 95 percent of them are compounds of carbon, one realizes that carbon is unique. Why are there so many carbon compounds? It turns out that atoms of carbon are quite remarkable in a number of ways.

Carbon atoms form very strong bonds with other carbon atoms. The bonds are so strong that carbon can form long chains, some containing thousands of carbon atoms. (Carbon is the only element that can do this).

A carbon atom forms four bonds, therefore carbon not only can form long chains, but it also forms chains that have branches. It is a major reason why carbon compounds exhibit so much isomerism. The simple compound decane ($C_{10}H_{22}$), for example, has 75 different isomers .

Carbon atoms can be bonded by double or triple bonds as well as single bonds. This multiple bonding is much more prevalent with carbon than with any other element.

Carbon atoms can form rings of various sizes. The rings may be saturated or unsaturated. The unsaturated 6-membered ring known as the benzene ring is the basis for an entire subfield of "aromatic" organic chemistry.

Carbon atoms form strong bonds not only with other carbon atoms but also with atoms of other elements. In addition to hydrogen, many carbon compounds also contain oxygen. Nitrogen, sulfur, phosphorus, and the halogens also frequently occur in carbon compounds.

Various kinds of functional groups occur widely among carbon compounds, and many different kinds of isomers are possible.

Hydrocarbons

Compounds of carbon and hydrogen only are called hydrocarbons. These are the simplest compounds of organic chemistry. The most basic group of hydrocarbons are the alkanes, which contain only single bonds. The simplest member of the alkane series is methane, CH_4, the main component of natural gas. The names of some alkanes are listed in table. Alkanes sometimes have ring structures. Since a 4-carbon chain of the alkane series is called *butane,* a ring of 4 carbon atoms is called *cyclobutane.*

ALKANES			
Formula		Name	
CH_4	CH_4	methane	gases
C_2H_6	CH_3CH_3	ethane	
C_3H_8	$CH_3CH_2CH_3$	propane	
C_4H_{10}	$CH_3CH_2CH_2CH_3$	butane	
C_5H_{12}	$CH_3(CH_2)_3CH_3$	pentane	liquids
C_6H_{14}	$CH_3(CH_2)_4CH_3$	hexane	
C_7H_{16}	$CH_3(CH_2)_5CH_3$	heptane	
C_8H_{18}	$CH_3(CH_2)_6CH_3$	octane	
C_9H_{20}	$CH_3(CH_2)_7CH_3$	nonane	
$C_{10}H_{22}$	$CH_3(CH_2)_8CH_3$	decane	
$C_{11}H_{24}$	$CH_3(CH_2)_9CH_3$	undecane	
$C_{12}H_{26}$	$CH_3(CH_2)_{10}CH_3$	dodecane	
$C_{13}H_{28}$	$CH_3(CH_2)_{11}CH_3$	tridecane	
$C_{14}H_{30}$	$CH_3(CH_2)_{12}CH_3$	tetradecane	
$C_{15}H_{32}$	$CH_3(CH_2)_{13}CH_3$	pentadecane	
$C_{16}H_{34}$	$CH_3(CH_2)_{14}CH_3$	hexadecane	
$C_{17}H_{36}$	$CH_3(CH_2)_{15}CH_3$	heptadecane	

$C_{18}H_{38}$	$CH_3(CH_2)_{16}CH_3$	octadecane	solids
$C_{19}H_{40}$	$CH_3(CH_2)_{17}CH_3$	nonadecane	
$C_{20}H_{42}$	$CH_3(CH_2)_{18}CH_3$	eicosane	

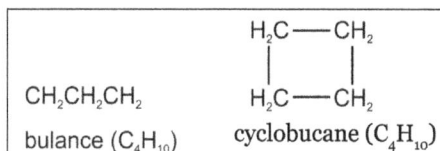

$CH_2CH_2CH_2$

bulance (C_4H_{10})

$$H_2C — CH_2$$
$$|\qquad|$$
$$H_2C — CH_2$$

cyclobucane (C_4H_{10})

Simple hydrocarbons that contain one or more double bonds are called *alkenes*. They are named like alkanes, but their names end in " –ene." The simplest alkene has two carbon atoms and is called *ethene*. A 3-carbon chain that has a double bond is called *propene*.

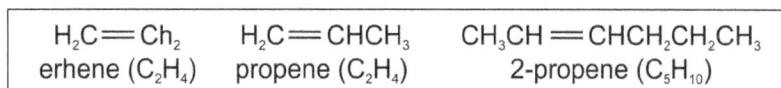

$H_2C = Ch_2$ $H_2C = CHCH_3$ $CH_3CH = CHCH_2CH_2CH_3$
erhene (C_2H_4) propene (C_2H_4) 2-propene (C_5H_{10})

A 5-carbon hydrocarbon chain with a double bond is called *pentene,* and if the double bond links the second and third carbons, it is 2- *pentene.* Like cycloalkanes, alkenes have the general formula C_nH_{2n}. Alkenes having ring structures are called *cycloalkenes.* A 5-carbon ring with a double bond is called *cyclopentene.*

Hydrocarbons that contain one or more triple bonds are called *alkynes,* and is the name ending is "–yne." A 2-carbon alkyne is therefore named *ethyne.* (However, the compound is often referred to by its common name, which is *acetylene*).

Compounds that contain double or triple bonds are said to be "unsaturated"—because they are not "saturated" with hydrogen atoms. Unsaturated compounds are reactive materials that readily add hydrogen when heated over a catalyst such as nickel. The reverse reaction also occurs. Heating ethane with steam is an important commercial process for making ethene (or ethylene). This is an important commercial process called "steam cracking."

When a 6-carbon ring contains 2 double bonds, it is called *cyclohexadiene,* but when it has 3 double bonds, it is not called cyclohexatriene; this is because a 6-carbon ring with three double bonds takes on a special kind of stability. The double bonds become completely conjugated and no longer behave as double bonds. The ring, known as a "benzene ring," is said to be aromatic.

The removal of a hydrogen atom from a hydrocarbon molecule leaves an alkyl group that readily attaches to a functional group, or forms a branch on a hydrocarbon chain. The groups are named after the corresponding hydrocarbons. For example, CH_3 – is named *methyl;* CH_3CH_2 –, *ethyl;* $CH_2 = CH–$, *ethenyl;* $CH_3CH_2CH_2$ –, *propyl;* and so on. A benzene ring from which a hydrogen atom has been removed is often referred to as a *phenyl.* The branched molecules shown here would be given names as follows:

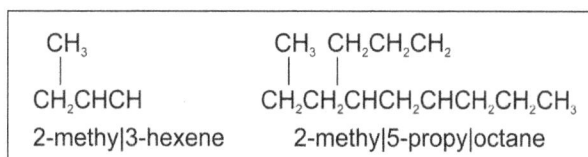

CH_3
$|$
CH_2CHCH

2-methy|3-hexene

$CH_3\ CH_2CH_2CH_2$
$|\qquad|$
$CH_2CH_2CHCH_2CHCH_2CH_2CH_3$

2-methy|5-propy|octane

Theoretically there is no limit to the length of hydrocarbon chains. Very large hydrocarbon molecules (polymers) have been made containing as many as 100,000 carbon atoms. However, such molecules are hard to make and very difficult to melt and to shape into useful products.

Hydrocarbons are obtained primarily from fossil fuels—especially petroleum and natural gas. Natural gas is a mixture that is largely methane mixed with varying amounts of ethane and other light hydrocarbons, while petroleum is a complex mixture of many different hydrocarbons. Coal, the other fossil fuel, is a much more complicated material from which many kinds of organic compounds, some of them hydrocarbons, can be obtained.

Functional Groups

Alkane molecules are rather unreactive (except for being very flammable), but alkenes react with many other substances. When a drop of bromine is added to an alkene, for example, the deep orange color of the bromine immediately disappears as the bromine adds across the double bond to form a dibromo derivative. The double bond is called a "functional group" because its presence in a molecule causes reactivity at that particular site. There are a dozen or so functional groups that appear frequently in organic compounds. The same molecule may contain several functional groups. Aspirin, for example, is both a carboxylic acid and an ester, and cholesterol is an alkene as well as an alcohol.

Isomerism

Isomers are molecules with the same molecular formula but different structures. There is only one structure for methane, ethane, or propane; but butane, C_4H_{10}, can have either of two different structure:

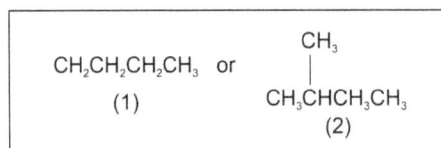

$$CH_2CH_2CH_2CH_3 \quad or \qquad \begin{matrix} CH_3 \\ | \\ CH_3CHCH_3CH_3 \end{matrix}$$
$$(1) \qquad\qquad\qquad (2)$$

The linear molecule (1) is called butane, or *normal* butane (*n* -butane), whereas the branched molecule (2) is methylpropane (rather than 2-methylpropane, as the methyl group has to be in a 2-position). If the methyl group of (2) were attached to a terminal carbon, the resultant molecule would be the same as (1). Methylpropane (2) is also called *iso* butane.

In a conjugated system, there are alternating double and single bonds, allowing electrons to flow back and forth. Molecules that contain such conjugated systems are said to be stabilized by "resonance." In the benzene ring every other bond is a double bond, all the way around the ring. This results in a special kind of stabilization called "aromaticity," in which the electrons are delocalized and free to travel all around the ring. Certain ring compounds, like benzene, that contain such a conjugated system of double and single bonds are described as "aromatic."

Pentane has 3 isomers: pentane (or n -pentane), methylbutane (or iso- pentane), and dimethyl propane (or neo pentane). Hexane has 5 isomers: hexane, 2-methylpentane, 3-methylpentane, 2,2-dimethylbutane, and 2,3-dimethylbutane. Heptane has 9 different isomers, octane has 18, nonane has 35, and decane has 75. An increase in the number of carbon atoms greatly increases the possibilities for isomerism. There are more than 4,000 isomers of $C_{15}H_{32}$ and more than 366,000 isomers of $C_{20}H_{42}$. The formula $C_{30}H_{62}$ has more than 4 billion. Of course, most of them have never been isolated as pure compounds (but could be, if there were any point in doing it).

For molecules other than hydrocarbons, still other kinds of isomers are possible. The simple formula C_2H_6O can represent ethyl alcohol or dimethyl ether; and C_3H_6O could stand for an alcohol, an ether, an aldehyde, or a ketone (among other things). The larger a molecule is, and the greater the variety of atoms and functional groups it contains, the more numerous its isomers.

There is still another kind of isomerism that stems from the existence of "right-" and "left-handed" molecules. It is sometimes referred to as optical isomerism because the molecules that make up a pair of these isomers usually differ only in the way they rotate plane polarized light.

Nomenclature

There are so many millions of organic compounds that simply finding names for them all is a major challenge. It was not until the late nineteenth century that chemists developed a logical system for naming organic compounds. Compounds had often been named according to their sources. The 1-carbon carboxylic acid, for example, was first obtained from ants, and so it was called formic acid. The 2-carbon acid was obtained from vinegar and was called acetic acid.

To bring some order to the naming process an international meeting was held in 1892 at Geneva, Switzerland. The group later became known as the International Union of Pure and Applied Chemistry (IUPAC). Its objective was to establish a naming process that would provide each compound with a unique and systematic name. An initial set of rules was adopted at that first meeting in Geneva, and IUPAC has continued that work. Its systematic naming rules are used by organic chemists all over the world. The names of the alkanes form the basis for the system, with functional groups usually being indicated with appropriate suffixes.

Future Sources of Organic Chemicals

Fossil fuels have been our primary natural source for many organic chemicals for more than a century, but our fossil fuel resources are finite, and they are being rapidly depleted (especially oil and gas). What will be our sources of organic materials in the future? Since fossil fuels are non-renewable resources, it is believed that the twenty-first century will see a shift toward greater dependence on renewable raw materials. The largest U.S. chemical company has a goal of becoming 25 percent based on renewable resources by 2010. It is already producing 1,3-propanediol from cornstarch using a gene-tailored E. coli bacterium. This diol is used in Du Pont's fiber Sorona, which is said to combine the best features of both polyester and nylon fibers. Succinic acid and polyhydroxybutyrate are also obtainable from renewable crops, and the list of such renewable raw materials is destined to grow. For example, ethylene (or ethene), $CH_2 = CH_2$, which is a highly important commercial chemical used in making many industrial chemicals and polymers, is presently made by steam cracking of ethane obtained from oil or natural gas; but ethylene can also be made by dehydration of ethyl alcohol made by fermentation of sugar. Efforts are even being made to use biowaste materials, such as corn husks, nutshells, and wood chips as industrial raw materials.

Analytical Tools

Organic chemists often need to examine products for identification, purity analysis, or structure determination. There are some marvelous tools available to help them do these things. Chromatography, spectroscopy, and crystallography are especially widely used in organic chemistry.

Column chromatography, gas chromatography, and liquid chromatography are all important methods for separating mixtures of organic compounds. Spectroscopic tools include ultraviolet (UV), infrared (IR), nuclear magnetic resonance (NMR), and mass spectroscopy (MS), each capable of providing a different kind of information about an organic compound. Although it is limited to substances that can be prepared as pure crystals, x-ray crystallography is probably the ultimate tool for determining molecular structure.

IUPAC Nomenclature of Organic Chemistry

In chemical nomenclature, the IUPAC nomenclature of organic chemistry is a systematic method of naming organic chemical compounds as recommended by the International Union of Pure and Applied Chemistry (IUPAC). It is published in the *Nomenclature of Organic Chemistry* (informally called the Blue Book). Ideally, every possible organic compound should have a name from which an unambiguous structural formula can be created. There is also an IUPAC nomenclature of inorganic chemistry.

To avoid long and tedious names in normal communication, the official IUPAC naming recommendations are not always followed in practice, except when it is necessary to give an unambiguous and absolute definition to a compound. IUPAC names can sometimes be simpler than older names, as with ethanol, instead of ethyl alcohol. For relatively simple molecules they can be more easily understood than non-systematic names, which must be learnt or looked up. However, the common or trivial name is often substantially shorter and clearer, and so preferred. These non-systematic names are often derived from an original source of the compound. In addition, very long names may be less clear than structural formulae.

Basic Principles

In chemistry, a number of prefixes, suffixes and infixes are used to describe the type and position of functional groups in the compound.

The steps for naming an organic compound are:

- Identification of the parent hydrocarbon chain. This chain must obey the following rules, in order of precedence:

    ○ It should have the maximum number of substituents of the suffix functional group. By suffix, it is meant that the parent functional group should have a suffix, unlike halogen substituents. If more than one functional group is present, the one with highest precedence should be used.

    ○ It should have the maximum number of multiple bonds.

- It should have the maximum number of single bonds.

- It should have the maximum length.

- Identification of the parent functional group, if any, with the highest order of precedence.

- Identification of the side-chains. Side chains are the carbon chains that are not in the parent chain, but are branched off from it.

- Identification of the remaining functional groups, if any, and naming them by their ionic prefixes (such as hydroxy for -OH, oxy for =O, oxyalkane for O-R, etc.). Different side-chains and functional groups will be grouped together in alphabetical order. (The prefixes di-, tri-, etc. are not taken into consideration for grouping alphabetically. For example, ethyl comes before dihydroxy or dimethyl, as the "e" in "ethyl" precedes the "h" in "dihydroxy" and the "m" in "dimethyl" alphabetically. The "di" is not considered in either case). When both side chains and secondary functional groups are present, they should be written mixed together in one group rather than in two separate groups.

- Identification of double/triple bonds.

- Numbering of the chain. This is done by first numbering the chain in both directions (left to right and right to left), and then choosing the numbering which follows these rules, in order of precedence:

  - Has the lowest-numbered locant (or locants) for the suffix functional group. Locants are the numbers on the carbons to which the substituent is directly attached.

  - Has the lowest-numbered locants for multiple bonds (The locant of a multiple bond is the number of the adjacent carbon with a lower number).

  - Has the lowest-numbered locants for prefixes.

- Numbering of the various substituents and bonds with their locants. If there is more than one of the same type of substituent/double bond, a prefix is added showing how many there are (di – 2 tri – 3 tetra – 4 then as for the number of carbons below with 'a' added).

The numbers for that type of side chain will be grouped in ascending order and written before the name of the side-chain. If there are two side-chains with the same alpha carbon, the number will be written twice. Example: 2,2,3-trimethyl-. If there are both double bonds and triple bonds, "en" (double bond) is written before "yne" (triple bond). When the main functional group is a terminal functional group (a group which can exist only at the end of a chain, like formyl and carboxyl groups), there is no need to number it.

- Arrangement in this form: Group of side chains and secondary functional groups with numbers made in step 3 + prefix of parent hydrocarbon chain (eth, meth) + double/triple bonds with numbers (or "ane") + primary functional group suffix with numbers. Wherever it says "with numbers", it is understood that between the word and the numbers, the prefix(di-, tri-) is used.

- Adding of punctuation:

  - Commas are put between numbers (2 5 5 becomes 2,5,5).

○ Hyphens are put between a number and a letter (2 5 5 trimethylheptane becomes 2,5,5-trimethylheptane).

○ Successive words are merged into one word (trimethyl heptane becomes trimethylheptane).

The finalized name should look like this:

#,#-di<side chain>-#-<secondary functional group>-#-<side chain>-#,#,#-tri<secondary functional group><parent chain prefix><If all bonds are single bonds, use "ane">-#,#-di<double bonds>-#-<triple bonds>-#-<primary functional group>

The group secondary functional groups and side chains may not look the same as shown here, as the side chains and secondary functional groups are arranged alphabetically. The di- and tri- have been used just to show their usage. (di- after #,#, tri- after #,#,#, etc.)

Example

Here is a sample molecule with the parent carbons numbered:

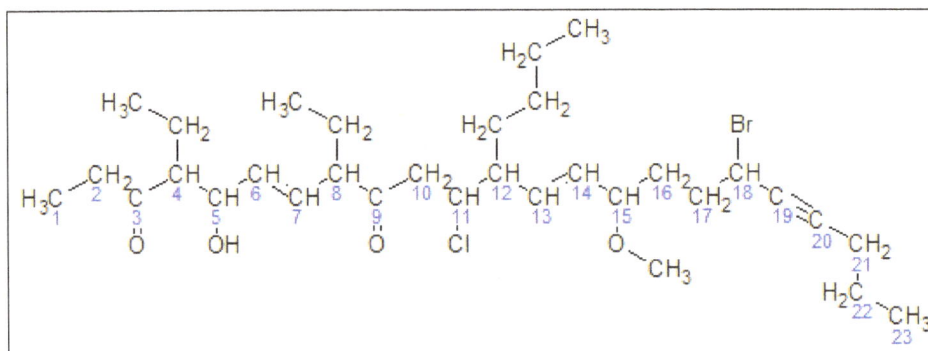

For simplicity, here is an image of the same molecule, where the hydrogens in the parent chain are removed and the carbons are shown by their numbers:

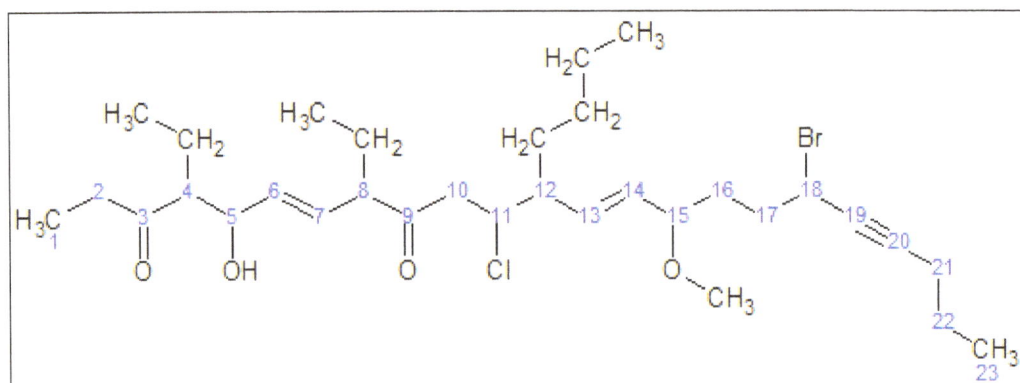

Now, following the above steps:

- The parent hydrocarbon chain has 23 carbons. It is called tricosa.

- The functional groups with the highest precedence are the two ketone groups.

- ◦ The groups are on carbon atoms 3 and 9. As there are two, we write 3,9-dione.

- ◦ The numbering of the molecule is based on the ketone groups. When numbering from left to right, the ketone groups are numbered 3 and 9. When numbering from right to left, the ketone groups are numbered 15 and 21. 3 is less than 15, therefore the ketones are numbered 3 and 9. The smaller number is always used, not the sum of the constituents numbers.

- The side chains are: an ethyl- at carbon 4, an ethyl- at carbon 8, and a butyl- at carbon 12:

  - ◦ There are two ethyl- groups. They are combined to create, 4,8-diethyl.

  - ◦ The side chains are grouped like this: 12-butyl-4,8-diethyl. (But this is not necessarily the final grouping, as functional groups may be added in between to ensure all groups are listed alphabetically).

- The secondary functional groups are: a hydroxy- at carbon 5, a chloro- at carbon 11, a methoxy- at carbon 15, and a bromo- at carbon 18. Grouped with the side chains, this gives 18-bromo-12-butyl-11-chloro-4,8-diethyl-5-hydroxy-15-methoxy.

- There are two double bonds: one between carbons 6 and 7, and one between carbons 13 and 14. They would be called "6,13-diene", but the presence of alkynes switches it to 6,13-dien. There is one triple bond between carbon atoms 19 and 20. It will be called 19-yne.

- The arrangement (with punctuation) is: 18-bromo-12-butyl-11-chloro-4,8-diethyl-5-hydroxy-15-methoxytricosa-6,13-dien-19-yne-3,9-dione.

- Finally, due to cis-trans isomerism, we have to specify the relative orientation of functional groups around each double bond. For this example, we have (6*E*,13*E*).

The final name is (6*E*,13*E*)-18-bromo-12-butyl-11-chloro-4,8-diethyl-5-hydroxy-15-methoxytricosa-6,13-dien-19-yne-3,9-dione.

Hydrocarbons

Alkanes

Straight-chain alkanes take the suffix "-ane" and are prefixed depending on the number of carbon atoms in the chain, following standard rules. The first few are:

Number of carbons	1	2	3	4	5	6	7	8	9	10	11	12	13	14	15	16	17	18	19	20
Prefix	Meth	Eth	Prop	But	Pent	Hex	Hept	Oct	Non	Dec	Undec	Dodec	Tridec	Tetradec	Pentadec	Hexadec	Heptadec	Octadec	Nonadec	Eicos

For example, the simplest alkane is CH_4 methane, and the nine-carbon alkane $CH_3(CH_2)_7CH_3$ is named nonane. The names of the first four alkanes were derived from methanol, ether, propionic acid and butyric acid, respectively. The rest are named with a Greek numeric prefix, with the exceptions of nonane which has a Latin prefix, and undecane and tridecane which have mixed-language prefixes.

Cyclic alkanes are simply prefixed with "cyclo-": for example, C_4H_8 is cyclobutane (not to be confused with butene) and C_6H_{12} is cyclohexane.

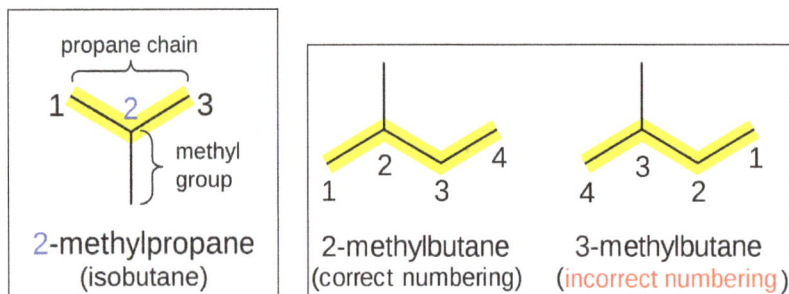

2-methylpropane
(isobutane)

2-methylbutane
(correct numbering)

3-methylbutane
(incorrect numbering)

Branched alkanes are named as a straight-chain alkane with attached alkyl groups. They are prefixed with a number indicating the carbon the group is attached to, counting from the end of the alkane chain. For example, $(CH_3)_2CHCH_3$, commonly known as isobutane, is treated as a propane chain with a methyl group bonded to the middle (2) carbon, and given the systematic name 2-methylpropane. However, although the name 2-methylpropane *could* be used, it is easier and more logical to call it simply methylpropane – the methyl group could not possibly occur on any of the other carbon atoms (that would lengthen the chain and result in butane, not propane) and therefore the use of the number "2" is unnecessary.

If there is ambiguity in the position of the substituent, depending on which end of the alkane chain is counted as "1", then numbering is chosen so that the smaller number is used. For example, $(CH_3)_2CHCH_2CH_3$ (isopentane) is named 2-methylbutane, not 3-methylbutane.

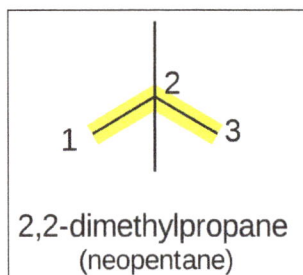

2,2-dimethylpropane
(neopentane)

If there are multiple side-branches of the same size alkyl group, their positions are separated by commas and the group prefixed with di-, tri-, tetra-, etc., depending on the number of branches. For example, $C(CH_3)_4$ (neopentane) is named 2,2-dimethylpropane. If there are different groups, they are added in alphabetical order, separated by commas or hyphens:. The longest possible main alkane chain is used; therefore 3-ethyl-4-methylhexane instead of 2,3-diethylpentane, even though these describe equivalent structures. The di-, tri- etc. prefixes are ignored for the purpose of alphabetical ordering of side chains (e.g. 3-ethyl-2,4-dimethylpentane, not 2,4-dimethyl-3-ethylpentane).

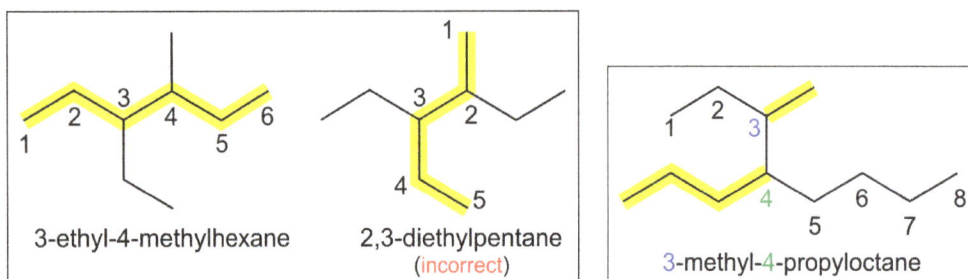

3-ethyl-4-methylhexane

2,3-diethylpentane
(incorrect)

3-methyl-4-propyloctane

Alkenes

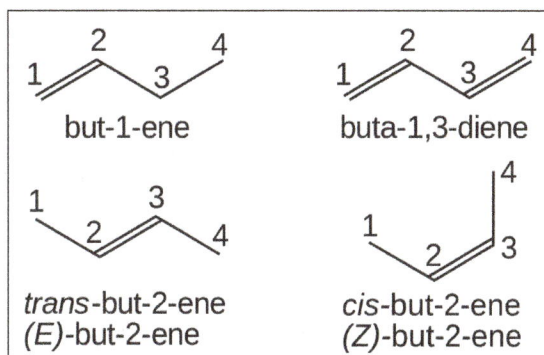

but-1-ene

buta-1,3-diene

trans-but-2-ene
(*E*)-but-2-ene

cis-but-2-ene
(*Z*)-but-2-ene

Alkenes are named for their parent alkane chain with the suffix "-ene" and an infixed number indicating the position of the carbon with the lower number for each double bond in the chain: $CH_2=CHCH_2CH_3$ is but-1-ene. Multiple double bonds take the form -diene, -triene, etc., with the size prefix of the chain taking an extra "a": $CH_2=CHCH=CH_2$ is buta-1,3-diene. Simple cis and trans isomers may be indicated with a prefixed *cis*- or *trans*-: *cis*-but-2-ene, *trans*-but-2-ene. However, *cis*- and *trans*- are *relative* descriptors. It is IUPAC convention to describe all alkenes using *absolute* descriptors of *Z*- (same side) and *E*- (opposite) with the Cahn–Ingold–Prelog priority rules.

Alkynes

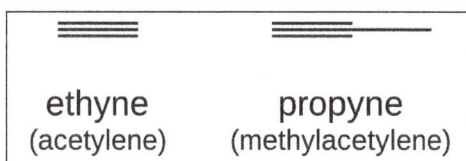

ethyne
(acetylene)

propyne
(methylacetylene)

Alkynes are named using the same system, with the suffix "-yne" indicating a triple bond: ethyne (acetylene), propyne (methylacetylene).

Functional Groups

Haloalkanes and Haloarenes

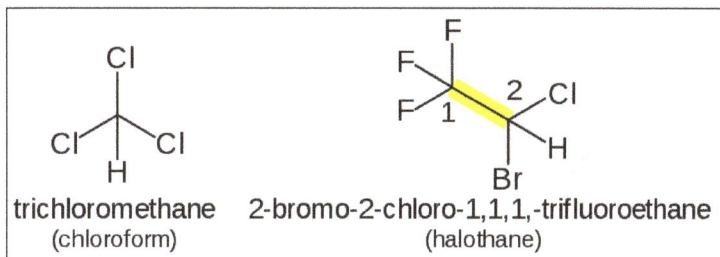

trichloromethane
(chloroform)

2-bromo-2-chloro-1,1,1-trifluoroethane
(halothane)

In Haloalkanes and Haloarenes (R-X), Halogen functional groups are prefixed with the bonding position and take the form of fluoro-, chloro-, bromo-, iodo-, etc., depending on the halogen. Multiple groups are dichloro-, trichloro-, etc., and dissimilar groups are ordered alphabetically as before. For example, $CHCl_3$ (chloroform) is trichloromethane. The anesthetic Halothane ($CF_3CHBrCl$) is 2-bromo-2-chloro-1,1,1-trifluoroethane.

Alcohols

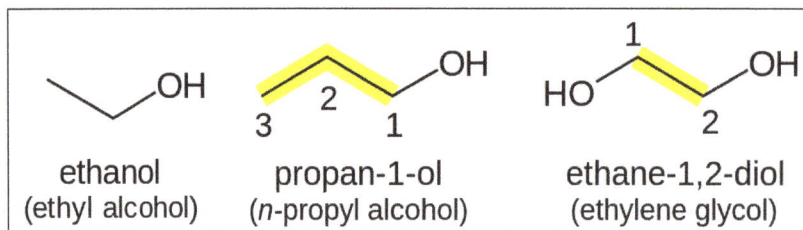

ethanol	propan-1-ol	ethane-1,2-diol
(ethyl alcohol)	(*n*-propyl alcohol)	(ethylene glycol)

Alcohols (R-OH) take the suffix "-ol" with an infix numerical bonding position: $CH_3CH_2CH_2OH$ is propan-1-ol. The suffixes -diol, -triol, -tetraol, etc., are used for multiple -OH groups: Ethylene glycol CH_2OHCH_2OH is ethane-1,2-diol.

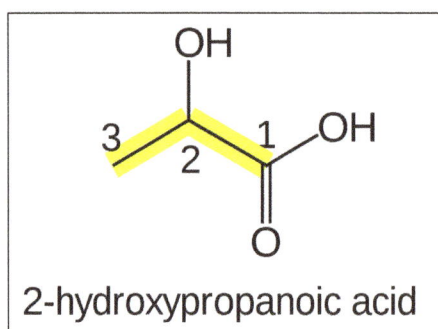

2-hydroxypropanoic acid

If higher precedence functional groups are present, the prefix "hydroxy" is used with the bonding position: $CH_3CHOHCOOH$ is 2-hydroxypropanoic acid.

Ethers

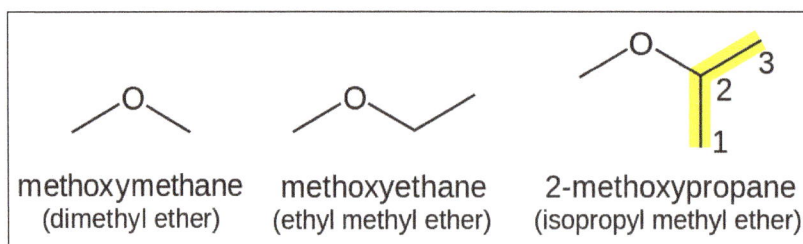

methoxymethane	methoxyethane	2-methoxypropane
(dimethyl ether)	(ethyl methyl ether)	(isopropyl methyl ether)

Ethers (R-O-R) consist of an oxygen atom between the two attached carbon chains. The shorter of the two chains becomes the first part of the name with the -ane suffix changed to -oxy, and the longer alkane chain becomes the suffix of the name of the ether. Thus, CH_3OCH_3 is methoxymethane, and $CH_3OCH_2CH_3$ is methoxyethane (*not* ethoxymethane). If the oxygen is not attached to the end of the main alkane chain, then the whole shorter alkyl-plus-ether group is treated as a side-chain and prefixed with its bonding position on the main chain. Thus $CH_3OCH(CH_3)_2$ is 2-methoxypropane.

Alternatively, an ether chain can be named as an alkane in which one carbon is replaced by an oxygen, a replacement denoted by the prefix "oxa". For example, $CH_3OCH_2CH_3$ could also be called 2-oxabutane, and an epoxide could be called oxacyclopropane. This method is especially useful when both groups attached to the oxygen atom are complex.

Aldehydes

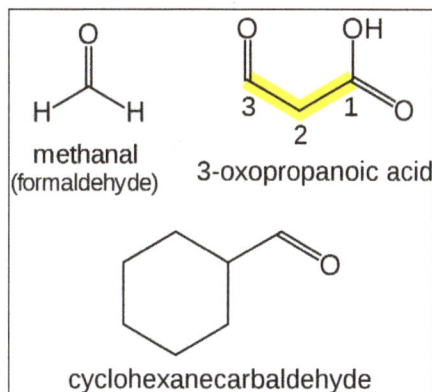

methanal
(formaldehyde)

3-oxopropanoic acid

cyclohexanecarbaldehyde

Aldehydes (R-CHO) take the suffix "-al". If other functional groups are present, the chain is numbered such that the aldehyde carbon is in the "1" position, unless functional groups of higher precedence are present.

If a prefix form is required, "oxo-" is used (as for ketones), with the position number indicating the end of a chain: $CHOCH_2COOH$ is 3-oxopropanoic acid. If the carbon in the carbonyl group cannot be included in the attached chain (for instance in the case of cyclic aldehydes), the prefix "formyl-" or the suffix "-carbaldehyde" is used: $C_6H_{11}CHO$ is cyclohexanecarbaldehyde. If an aldehyde is attached to a benzene and is the main functional group, the suffix becomes benzaldehyde.

Ketones

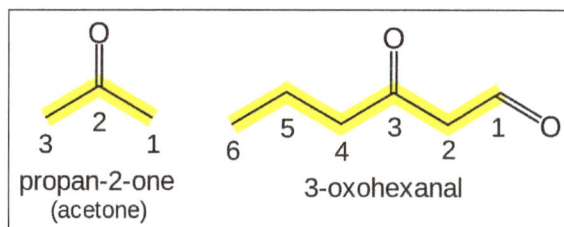

propan-2-one
(acetone)

3-oxohexanal

In general ketones (R-CO-R) take the suffix "-one" (pronounced *own*, not *won*) with an infix position number: $CH_3CH_2CH_2COCH_3$ is pentan-2-one. If a higher precedence suffix is in use, the prefix "oxo-" is used: $CH_3CH_2CH_2COCH_2CHO$ is 3-oxohexanal.

Carboxylic Acids

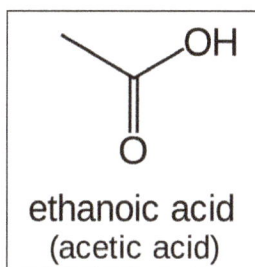

ethanoic acid
(acetic acid)

In general, carboxylic acids are named with the suffix -*oic acid* (etymologically a back-formation from benzoic acid). As with aldehydes, the carboxyl functional group must take the "1" position on the main chain and so the locant need not be stated. For example, CH_3-CH(OH)-COOH (lactic acid) is named 2-hydroxypropanoic acid with no "1" stated. Some traditional names for common carboxylic acids (such as acetic acid) are in such widespread use that they are retained in IUPAC nomenclature, though systematic names like ethanoic acid are also used. Carboxylic acids attached to a benzene ring are structural analogs of benzoic acid (Ph-COOH) and are named as one of its derivatives.

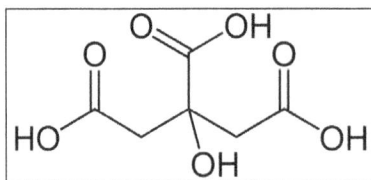
Citric Acid.

If there are multiple carboxyl groups on the same parent chain, multiplying prefixes are used: Malonic acid, $CH_2(COOH)_2$, is systematically named propanedioic acid. Alternatively, the suffix "-carboxylic acid" can be used, combined with a multiplying prefix if necessary – mellitic acid is benzenehexacarboxylic acid, for example. In the latter case, the carbon atom(s) in the carboxyl group(s) do *not* count as being part of the main chain, a rule that also applies to the prefix form "carboxy-". Citric acid serves as an example: it is formally named 2-hydroxypropane-1,2,3-tricarboxylic acid rather than 3-carboxy-3-hydroxypentanedioic acid.

Carboxylates

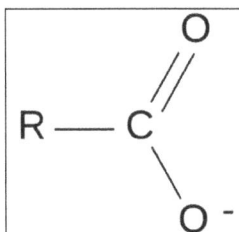

Salts of carboxylic acids are named following the usual cation-then-anion conventions used for ionic compounds in both IUPAC and common nomenclature systems. The name of the carboxylate anion is derived from that of the parent acid by replacing the "–oic acid" ending with "–oate." For example, $C_6H_5CO_2Na$, the sodium salt of benzoic acid (C_6H_5COOH), is called sodium benzoate. Where an acid has both a systematic and a common name (like CH_3COOH, for example, which is known as both acetic acid and as ethanoic acid), its salts can be named from either parent name. Thus, CH_3CO_2K can be named as potassium acetate or as potassium ethanoate.

Esters

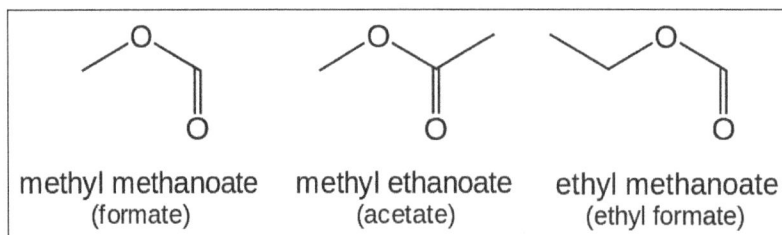
methyl methanoate   methyl ethanoate   ethyl methanoate
(formate)   (acetate)   (ethyl formate)

Esters (R-CO-O-R') are named as alkyl derivatives of carboxylic acids. The alkyl (R') group is named first. The R-CO-O part is then named as a separate word based on the carboxylic acid name, with the ending changed from *-oic acid* to *-oate*. For example, $CH_3CH_2CH_2CH_2COOCH_3$ is *methyl pentanoate*, and $(CH_3)_2CHCH_2CH_2COOCH_2CH_3$ is *ethyl 4-methylpentanoate*. For esters such as ethyl acetate ($CH_3COOCH_2CH_3$), ethyl formate ($HCOOCH_2CH_3$) or dimethyl phthalate that are based on common acids, IUPAC recommends use of these established names, called retained names. The *-oate* changes to *-ate*. Some simple examples, named both ways, are shown in the figure above.

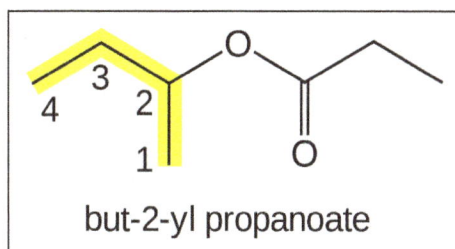

but-2-yl propanoate

If the alkyl group is not attached at the end of the chain, the bond position to the ester group is infixed before "-yl": $CH_3CH_2CH(CH_3)OOCCH_2CH_3$ may be called but-2-yl propanoate or but-2-yl propionate.

Acyl Groups

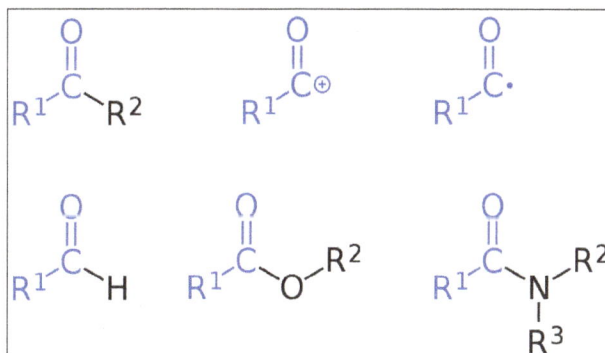

Acyl groups are named by stripping the -ic acid of the corresponding carboxylic acid and replacing it with -yl. For Example, CH3CO-R is called Ethanoyl-R.

Acyl Halides

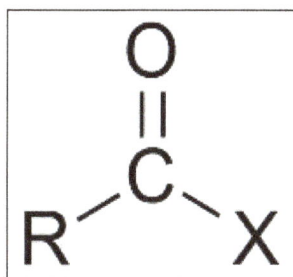

Simply add the name of the attached halide to the end of the acyl group. For example, CH_3COCl is Ethanoyl Chloride.

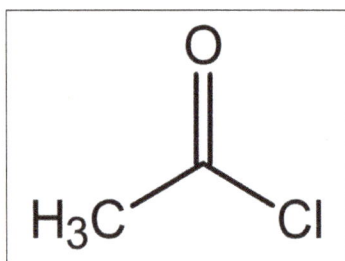

Acid Anhydrides

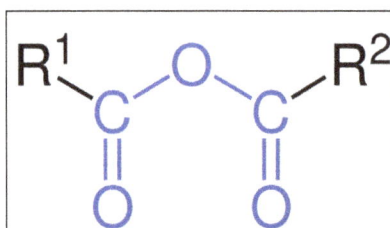

If both acyl groups are the same, then the name of the carboxylic acid with the word acid replaced with anhydride. If the acyl groups are different, then they are named in alphabetical order in the same way, with anhydride replacing acid. For example, $CH_3CO\text{-}O\text{-}OCCH_3$ is called Ethanoic Anhydride.

Amines

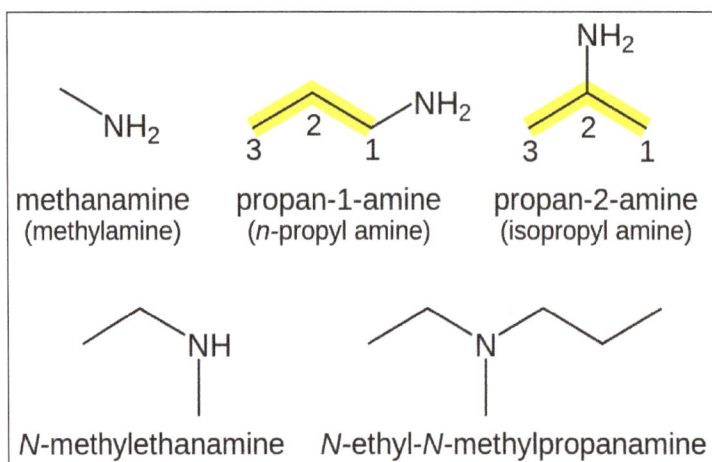

Amines ($R\text{-}NH_2$) are named for the attached alkane chain with the suffix "-amine" (e.g. CH_3NH_2 methanamine). If necessary, the bonding position is infixed: $CH_3CH_2CH_2NH_2$ propan-1-amine, $CH_3CHNH_2CH_3$ propan-2-amine. The prefix form is "amino-".

For secondary amines (of the form R-NH-R), the longest carbon chain attached to the nitrogen atom becomes the primary name of the amine; the other chain is prefixed as an alkyl group with location prefix given as an italic N: $CH_3NHCH_2CH_3$ is N-methylethanamine. Tertiary amines (R-NR-R) are treated similarly: $CH_3CH_2N(CH_3)CH_2CH_2CH_3$ is N-ethyl-N-methylpropanamine. Again, the substituent groups are ordered alphabetically.

Amides

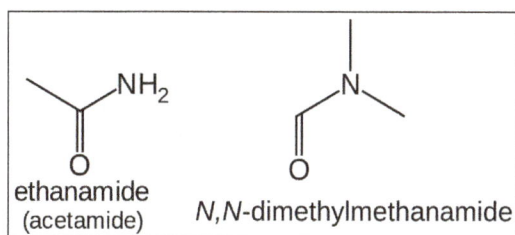

ethanamide
(acetamide)

N,N-dimethylmethanamide

Amides (R-CO-NH$_2$) take the suffix "-amide", or "-carboxamide" if the carbon in the amide group cannot be included in the main chain. The prefix form is both "carbamoyl-" and "amido-".

Amides that have additional substituents on the nitrogen are treated similarly to the case of amines: they are ordered alphabetically with the location prefix N: HCON(CH$_3$)$_2$ is N,N-dimethylmethanamide.

Nitriles

Nitriles (RCN) are named by adding the suffix -nitrile to the longest hydrocarbon chain (including the carbon of the cyano group). It can also be named by replacing the -oic acid of their corresponding carboxylic acids with -onitrile. Functional class IUPAC nomenclature may also be used in the form of alkyl cyanides. For example, CH$_3$CH$_2$CH$_2$CH$_2$CN is called pentanenitrile or butyl cyanide.

Cyclic Compounds

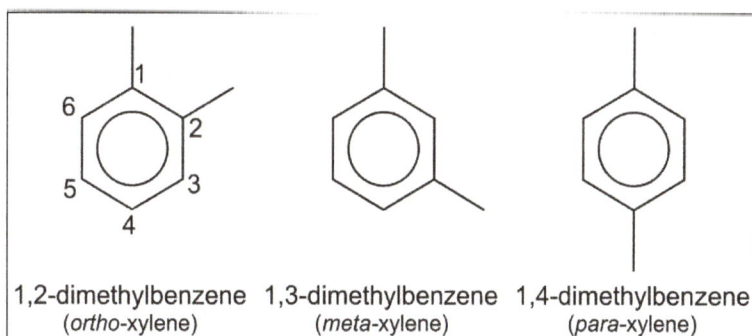

1,2-dimethylbenzene   1,3-dimethylbenzene   1,4-dimethylbenzene
(ortho-xylene)   (meta-xylene)   (para-xylene)

Cycloalkanes and aromatic compounds can be treated as the main parent chain of the compound, in which case the positions of substituents are numbered around the ring structure. For example, the three isomers of xylene CH$_3$C$_6$H$_4$CH$_3$, commonly the ortho-, meta-, and para- forms, are 1,2-dimethylbenzene, 1,3-dimethylbenzene, and 1,4-dimethylbenzene. The cyclic structures can also be treated as functional groups themselves, in which case they take the prefix "cycloalkyl-" (e.g. "cyclohexyl-") or for benzene, "phenyl-".

The IUPAC nomenclature scheme becomes rapidly more elaborate for more complex cyclic structures, with notation for compounds containing conjoined rings, and many common names such as phenol being accepted as base names for compounds derived from them.

Order of Precedence of Groups

When compounds contain more than one functional group, the order of precedence determines which groups are named with prefix or suffix forms. The table below shows common groups in decreasing order of precedence. The highest-precedence group takes the suffix, with all others taking the prefix form. However, double and triple bonds only take suffix form (-en and -yn) and are used with other suffixes.

Prefixed substituents are ordered alphabetically (excluding any modifiers such as di-, tri-, etc.), e.g. chlorofluoromethane, *not* fluorochloromethane. If there are multiple functional groups of the same type, either prefixed or suffixed, the position numbers are ordered numerically (thus ethane-1,2-diol, *not* ethane-2,1-diol.) The *N* position indicator for amines and amides comes before "1", e.g. $CH_3CH(CH_3)CH_2NH(CH_3)$ is *N*,2-dimethylpropanamine.

Priority	Functional group	Formula	Prefix	Suffix
1	Cations e.g. Ammonium	NH_4^+	-onio- ammonio-	-onium -ammonium

2	Carboxylic acids
	
	  Carbothioic *S*-acids
	
	  Carboselenoic *Se*-acids
	
	  Sulfonic acids
	
	  Sulfinic acids
		−COOH
		−COSH
		−COSeH
		−SO$_3$H
		−SO$_2$H
			carboxy-
			sulfanylcarbonyl-
			selanylcarbonyl-
			sulfo-
			sulfino-
				-oic acid*
				-thioic *S*-acid*
				-selenoic *Se*-acid*
				-sulfonic acid
				-sulfinic acid

3	*Carboxylic acid derivatives*
	
	  Esters
	
	  Acyl halides
	
	  Amides
	
	  Imides
	
	  Amidines
		−COOR
		−COX
		−CONH$_2$
		−CON=C<
		−C(=NH)
		NH$_2$
			R-oxycarbonyl-
			halocarbonyl-
			carbamoyl-
			-imido-
			amidino-
				-R-oate
				-oyl halide*
				-amide*
				-imide*
				-amidine*

4	Nitriles Isocyanides	−CN −NC	cyano- isocyano-	-nitrile* isocyanide
5	Aldehydes Thioaldehydes	−CHO −CHS	formyl- thioformyl-	-al* -thial*

6	Ketones
	
	  Thioketones
	
	  Selones
	
	  Tellones
		=O
		=S
		=Se
		=Te
			oxo-
			sulfanylidene-
			selanylidene-
			tellanylidene-
				-one
				-thione
				-selone
				-tellone

7	Alcohols
	
	  Thiols
	
	  Selenols
	
	  Tellurols
		−OH
		−SH
		−SeH
		−TeH
			hydroxy-
			sulfanyl-
			selanyl-
			tellanyl-
				-ol
				-thiol
				-selenol
				-tellurol

| 8 | Hydroperoxides

Peroxols

Thioperoxols (Sulfenic acid)

Dithioperoxols | -OOH
-SOH
-SSH | hydroperoxy-
hydroxysulfanyl-
disulfanyl- | -peroxol
-SO-thioperoxol
-dithioperoxol |
| 9 | Amines
Imines
Hydrazines | $-NH_2$
$=NH$
$-NHNH_2$ | amino-
imino-
hydrazino- | -amine
-imine
-hydrazine |

The order of remaining functional groups is only needed for substituted benzene and hence is not mentioned here.

Common Nomenclature – Trivial Names

Common nomenclature uses the older names for some organic compounds instead of using the prefixes for the carbon skeleton above. The pattern can be seen below:

Number of carbons	Prefix as in new system	Common name for alcohol	Common name for aldehyde	Common name for acid	Common name for ketone
1	Meth-	Methyl alcohol (wood alcohol)	Formaldehyde	Formic acid	NA
2	Eth-	Ethyl alcohol (grain alcohol)	Acetaldehyde	Acetic acid(vinegar)	NA
3	Prop-	Propyl alcohol	Propionaldehyde	Propionic acid	Acetone/dimethyl ketone
4	But-	Butyl alcohol	Butyraldehyde	Butyric acid	Ethyl methyl ketone
5	Pent-	Amyl alcohol	Valeraldehyde	Valeric acid	• Metyl propyl ketone • Diethyl ketone
6	Hex-	Caproyl alcohol	Caproaldehyde	Caproic acid	• Butyl methyl ketone • Ethyl propyl ketone
7	Hept-	Enanthyl alcohol	Enanthaldehyde	Enanthoic acid	• Methyl pentyl ketone • Butyl Ethyl ketone • Dipropyl ketone
8	Oct-	Capryl alcohol	Caprylaldehyde	Caprylic acid	• Hexyl methyl ketone • Ethyl pentyl ketone • Butyl propyl ketone
9	Non-	Pelargonic alcohol	Pelargonalde-hyde	Pelargonic acid	• Heptyl methyl ketone • Ethyl hexyl ketone • Pentyl propyl ketone • Dibutyl ketone
10	Dec-	Capric alcohol	Capraldehyde	Capric acid	• Methyl octyl ketone • Ethyl heptyl ketone • Hexyl propyl ketone • Butyl pentyl ketone

11	Undec-	-	-	-	
12	Dodec-	Lauryl alcohol	Lauraldehyde	Lauric acid	
13	Tridec-	-	-	-	
14	Tetradec-	Myristyl alcohol	Myristaldehyde	Myristic acid	
15	Pentadec-	-	-	-	
16	Hexadec-	Cetyl alcohol Palmityl alcohol	Palmitaldehyde	Palmitic acid	
17	Heptadec-	-	-	Margaric acid	
18	Octadec-	Stearyl alcohol	Stearaldehyde	Stearic acid	
19	Nonadec-	-	-	-	
20	Eicos-	Arachidyl alcohol	-	Arachidic acid	
21	Henicos-	-	-	-	
22	Docos-	Behenyl alcohol	-	Behenic acid	
23	Tricos-	-	-	-	
24	Tetracos-	Lignoceryl alcohol	-	Lignoceric acid	
25	Pentacos-	-	-	-	The same pattern continues
26	Hexacos-	Ceryl alcohol	-	Cerotic acid	
27	Heptacos-	-	-	-	
28	Octacos-	Montanyl alcohol	-	Montanic acid	
29	Nonacos-	-	-	-	
30	Triacont-	Melissyl alcohol	-	Melissic acid	
31	Hentriacont-	-	-	-	
32	Dotriacont-	Lacceryl alcohol	-	Lacceroic acid	
33	Tritriacont-	Psyllic alcohol	-	Psyllic acid	
34	Tetratriacont-	Geddyl alcohol	-	Geddic acid	
35	Pentatriacont-	-	-	Ceroplastic acid	
36	Hexatriacont-	-	-	-	
37	Heptatriacont-	-	-	-	
38	Octatriacont-	-	-	-	
39	Nonatriacont-	-	-	-	
40	Tetracont-	-	-	-	

Ketones

Common names for ketones can be derived by naming the two alkyl or aryl groups bonded to the carbonyl group as separate words followed by the word *ketone*.

- Acetone,

- Acetophenone,

- Benzophenone,

- Ethyl isopropyl ketone,

- Diethyl ketone.

The first three of the names shown above are still considered to be acceptable IUPAC names.

Aldehydes

The common name for an aldehyde is derived from the common name of the corresponding carboxylic acid by dropping the word *acid* and changing the suffix from -ic or -oic to -aldehyde.

- Formaldehyde,

- Acetaldehyde.

Ions

The IUPAC nomenclature also provides rules for naming ions.

Hydron

Hydron is a generic term for hydrogen cation; protons, deuterons and tritons are all hydrons. The Hydrons are not found in heavier isotopes, however.

Parent Hydride Cations

Simple cations formed by adding a hydron to a hydride of a halogen, chalcogen or pnictogen are named by adding the suffix "-onium" to the element's root: H_4N^+ is ammonium, H_3O^+ is oxonium, and H_2F^+ is fluoronium. Ammonium was adopted instead of nitronium, which commonly refers to NO_2^+.

If the cationic center of the hydride is not a halogen, chalcogen or pnictogen then the suffix "-ium" is added to the name of the neutral hydride after dropping any final 'e'. H_5C^+ is methanium, $HO-(O^+)-H_2$ is dioxidanium (HO-OH is dioxidane), and $H_2N-(N^+)-H_3$ is diazanium (H_2N-NH_2 is diazane).

Cations and Substitution

The above cations except for methanium are not, strictly speaking, organic, since they do not contain carbon. However, many organic cations are obtained by substituting another element or some functional group for a hydrogen.

The name of each substitution is prefixed to the hydride cation name. If many substitutions by the same functional group occur, then the number is indicated by prefixing with "di-", "tri-" as with halogenation. $(CH_3)_3O^+$ is trimethyloxonium. $CH_3F_3N^+$ is trifluoromethylammonium.

Organic Compound

An organic compound is a member of a class of chemicals containing carbon atoms linked to each other and to other atoms by covalent bonds and found in the cells of living organisms. Hydrogen, oxygen and nitrogen are typical elements that make up organic compounds in addition to carbon. Traces of other elements such as sulfur, phosphorous, iron and copper may also be present when needed for specific organic chemical reactions. The main groups of organic compounds are hydrocarbons, lipids, proteins and nucleic acids.

Organic compounds always contain carbon along with other elements that are needed for living organisms to function. Carbon is the key element because it has four electrons in an outer electron shell that can hold eight electrons. As a result, it can form many types of bonds with other carbon atoms and elements such as hydrogen, oxygen and nitrogen. Hydrocarbons and proteins are good examples of organic molecules that can form long chains and complex structures. The organic compounds made up of these molecules are the basis for chemical reactions in the cells of plants and animals – reactions that provide the energy for finding food, for reproduction and for all the other processes necessary for life.

Characteristics of Organic Compounds

The four types of organic compounds are hydrocarbons, lipids, proteins and nucleic acids, and they perform different functions in a living cell. While many organic compounds are not polar molecules and therefore don't dissolve well in the water of a cell, they often dissolve in other organic compounds. For example, while carbohydrates such as sugar are slightly polar and dissolve in water, fats do not. But fats dissolve in other organic solvents such as ethers. When in solution, the four types of organic molecules interact and form new compounds as they come into contact within living tissue.

Organic compounds range from simple substances with molecules made up of a few atoms of only two elements to long, complex polymers whose molecules include many elements. Hydrocarbons for example, are made up of only carbon and hydrogen. They can form simple molecules or long chains of atoms and are used for cell structure and as basic building blocks for more complex molecules.

Lipids are fats and similar materials that are made up of carbon, hydrogen and oxygen. They help form cell walls and membranes and are a major component of food. Proteins are made up of carbon, hydrogen, oxygen and nitrogen, and they have two major functions in cells. They form part of the cell and organ structures, but they are also enzymes, hormones and other organic chemicals that take part in chemical reactions to produce the materials essential for life.

Nucleic acids are made up of carbon, hydrogen, oxygen, nitrogen and phosphorous. As RNA and DNA, they store the instructions for chemical processes involving other proteins. They are the helix-shaped molecules of the genetic code. The four types of organic molecules are all based on carbon and a few other elements, but they have different properties.

Hydrocarbons

Hydrocarbons are the simplest organic compounds, and the simplest hydrocarbon is CH_4 or methane. The carbon atom shares electrons with four hydrogen atoms to complete its outer electron shell.

Instead of bonding with only hydrogen atoms, a carbon atom can share one or two of its outer shell electrons with another carbon atom, forming long chains. For example, butane, C_4H_{10}, is made up of a chain of four carbon atoms surrounded by 10 hydrogen atoms.

Lipids

A more complicated group of organic compounds are the lipids or fats. They include a hydrocarbon chain but also have a part where the chain bonds with oxygen. Organic compounds containing only carbon, hydrogen and oxygen are called carbohydrates.

Glycerol is an example of a simple lipid. Its chemical formula is $C_3H_8O_3$, and it has a chain of three carbon atoms with an oxygen atom bonded to each one. Glycerol is a building block that forms the base of many more complex lipids.

Proteins

Most proteins are very large molecules with complex structures that allow them to take on important roles in organic chemical reactions. In such reactions, parts of the proteins break apart, are rearranged or join with new chains. Even the simplest proteins have long chains and many subsections.

For example, 3-amino-2-butanol has the chemical formula $C_4H_{11}NO$, but it is really a sequence of hydrocarbon sections with a nitrogen and an oxygen atom attached. It is more clearly shown by the formula $CH_3CH(NH_2)CH(OH)CH_3$, and the amino acid is used in chemical reactions to produce other proteins.

Nucleic Acids

Nucleic acids form the basis of the genetic code of living cells and are long strings of repeating subunits. For nucleic acid deoxyribonucleic acid or DNA, for example, the molecules contain a phosphate group, a sugar and the repeating subunits cytosine, guanine, thymine and adenine. The part of a DNA molecule containing cytosine has a chemical formula $C_9H_{12}O_6N_3P$, and the sections containing different subunits form long polymer molecules located in the nucleus of cells.

Some organic compounds are the most complex molecules that exist, and they mirror the complexity of the chemical reactions that make life possible. Even with this complexity, the molecules are made up of relatively few elements, and all have carbon as a major component.

Functional Group

In organic chemistry, functional groups are specific substituents or moieties within molecules that are responsible for the characteristic chemical reactions of those molecules. The same functional group will undergo the same or similar chemical reaction(s) regardless of the size of the molecule it is a part of. This allows for systematic prediction of chemical reactions and behavior of chemical compounds and design of chemical syntheses. Furthermore, the reactivity of a functional group can be modified by other functional groups nearby. In organic synthesis, functional group inter-conversion is one of the basic types of transformations.

Functional groups are groups of one or more atoms of distinctive chemical properties no matter what they are attached to. The atoms of functional groups are linked to each other and to the rest of the molecule by covalent bonds. For repeating units of polymers, functional groups attach to their nonpolar core of carbon atoms and thus add chemical character to carbon chains. Functional groups can also be charged, e.g. in carboxylate salts ($-COO^-$), which turns the molecule into a polyatomic ion or a complex ion. Functional groups binding to a central atom in a coordination complex are called *ligands*. Complexation and solvation are also caused by specific interactions of functional groups. In the common rule of thumb "like dissolves like", it is the shared or mutually well-interacting functional groups which give rise to solubility. For example, sugar dissolves in water because both share the hydroxyl functional group ($-OH$) and hydroxyls interact strongly with each other. Plus, when functional groups are more electronegative than atoms they attach to, the functional groups will become polar, and the otherwise nonpolar molecules containing these functional groups become polar and so become soluble in some aqueous environment.

Combining the names of functional groups with the names of the parent alkanes generates what is termed a systematic nomenclature for naming organic compounds. In traditional nomenclature, the first carbon atom after the carbon that attaches to the functional group is called the alpha carbon; the second, beta carbon, the third, gamma carbon, etc. If there is another functional group at a carbon, it may be named with the Greek letter, e.g., the gamma-amine in gamma-aminobutyric acid is on the third carbon of the carbon chain attached to the carboxylic acid group. IUPAC conventions call for numeric labeling of the position, e.g. 4-aminobutanoic acid. In traditional names various qualifiers are used to label isomers, for example, isopropanol (IUPAC name: propan-2-ol) is an isomer of n-propanol (propan-1-ol).

Table of Common Functional Groups

The following is a list of common functional groups. In the formulas, the symbols R and R' usually denote an attached hydrogen, or a hydrocarbon side chain of any length, but may sometimes refer to any group of atoms.

Hydrocarbons

Functional groups, called hydrocarbyl, that contain only carbon and hydrogen, but vary in the number and order of double bonds. Each one differs in type (and scope) of reactivity.

Chemical class	Group	Formula	Structural Formulae	Prefix	Suffix	Example
Alkane	Alkyl	$R(CH_2)_nH$		alkyl-	-ane	Ethane

Alkene	Alkenyl	$R_2C=CR_2$		alkenyl-	-ene	
						Ethylene (*Ethene*)
Alkyne	Alkynyl	RC≡CR'		alkynyl-	-yne	Acetylene (*Ethyne*)
Benzene deriv-ative	Phenyl	RC_6H_5 RPh		phenyl-	-benzene	
						Cumene (*Isopropylbenzene*)

There are also a large number of branched or ring alkanes that have specific names, e.g., tert-butyl, bornyl, cyclohexyl, etc. Hydrocarbons may form charged structures: positively charged carbocations or negative carbanions. Carbocations are often named -*um*. Examples are tropylium and triphenylmethyl cations and the cyclopentadienyl anion.

Groups Containing Halogen

Haloalkanes are a class of molecule that is defined by a carbon–halogen bond. This bond can be relatively weak (in the case of an iodoalkane) or quite stable (as in the case of a fluoroalkane). In general, with the exception of fluorinated compounds, haloalkanes readily undergo nucleophilic substitution reactions or elimination reactions. The substitution on the carbon, the acidity of an adjacent proton, the solvent conditions, etc. all can influence the outcome of the reactivity.

Chemical class	Group	Formula	Structural Formula	Prefix	Suffix	Example
haloalkane	halo	RX	R–X	halo-	alkyl halide	Chloroethane (*Ethyl chloride*)
fluoroalkane	fluoro	RF	R–F	fluoro-	alkyl fluoride	Fluoromethane (*Methyl fluoride*)
chloroalkane	chloro	RCl	R–Cl	chloro-	alkyl chloride	Chloromethane (*Methyl chloride*)

bromoalkane	bromo	RBr	R–Br	bromo-	alkyl bromide	Bromomethane *(Methyl bromide)*
iodoalkane	iodo	RI	R–I	iodo-	alkyl iodide	Iodomethane *(Methyl iodide)*

Groups Containing Oxygen

Compounds that contain C-O bonds each possess differing reactivity based upon the location and hybridization of the C-O bond, owing to the electron-withdrawing effect of sp-hybridized oxygen (carbonyl groups) and the donating effects of sp^2-hybridized oxygen (alcohol groups).

Chemical class	Group	Formula	Structural Formula	Prefix	Suffix	Example
Alcohol	Hydroxyl	ROH	R–O–H	hydroxy-	-ol	Methanol
Ketone	Carbonyl	RCOR'	R–CO–R'	-oyl- (-COR') or oxo- (=O)	-one	Butanone *(Methyl ethyl ketone)*
Aldehyde	Aldehyde	RCHO	R–CO–H	formyl- (-COH) or oxo- (=O)	-al	Acetaldehyde *(Ethanal)*
Acyl halide	Halo-formyl	RCOX	R–CO–X	carbonofluoridoyl- carbonochloridoyl- carbonobromidoyl- carbonoiodidoyl-	-oyl halide	Acetyl chloride *(Ethanoyl chloride)*

Carbonate	Carbonate ester	ROCOOR'		(alkoxycarbonyl)oxy-	alkyl carbonate	Triphosgene (bis(trichloromethyl) carbonate)
Carboxylate	Carboxylate	RCOO⁻		carboxy-	-oate	Sodium acetate (Sodium ethanoate)
Carboxylic acid	Carboxyl	RCOOH		carboxy-	-oic acid	Acetic acid (Ethanoic acid)
Ester	Carboalkoxy	RCOOR'		alkanoyloxy- or alkoxycarbonyl	alkyl alkanoate	Ethyl butyrate (Ethyl butanoate)
Methoxy	Methoxy	ROCH₃		methoxy-		Anisole (Methoxybenzene)
Hydroperoxide	Hydroperoxy	ROOH		hydroperoxy-	alkyl hydroperoxide	tert-Butyl hydroperoxide
Peroxide	Peroxy	ROOR'		peroxy-	alkyl peroxide	Di-tert-butyl peroxide
Ether	Ether	ROR'		alkoxy-	alkyl ether	Diethyl ether (Ethoxyethane)
Hemiacetal	Hemiacetal	RCH(OR')(OH)		alkoxy -ol	-al alkyl hemiacetal	
Hemiketal	Hemiketal	RC(OR'')(OH)R'		alkoxy -ol	-one alkyl hemiketal	

Acetal	Acetal	RCH(OR') (OR")		dialkoxy-	-al dialkyl acetal	
Ketal (or Acetal)	Ketal (or Acetal)	RC(OR") (OR"')R'		dialkoxy-	-one dialkyl ketal	
Orthoester	Orthoe-ster	RC(OR') (OR") (OR"')		trialkoxy-		
Heterocycle	Methy-lenedioxy	PhOCOPh		methylenedioxy-	-dioxole	1,2-Methylenedi-oxybenzene (1,3-Benzodioxole)
Orthocar-bonate ester	Orthocar-bonate ester	C(OR) (OR')(OR") (OR")		tetralkoxy-	*tetraalkyl* orthocar-bonate	Tetramethoxy-methane
Organic acid anhy-dride	Carbox-ylic anhy-dride	R(CO) O(CO)R'			anhydride	Butyric anhydride

Groups Containing Nitrogen

Compounds that contain nitrogen in this category may contain C-O bonds, such as in the case of amides.

Chemical class	Group	Formula	Structural Formula	Prefix	Suffix	Example
Amide	Carboxam-ide	$RCONR_2$		carboxam-ido- or carbam-oyl-	-amide	Acetamide (*Ethanamide*)
Amines	Primary amine	RNH_2		amino-	-amine	Methylamine (*Methanamine*)
	Secondary amine	R_2NH		amino-	-amine	Dimethylamine

Amines	Tertiary amine	R_3N		amino-	-amine	 Trimethylamine
	4° ammonium ion	R_4N^+		ammonio-	-ammonium	 Choline
Imine	Primary ketimine	RC(=NH)R'		imino-	-imine	
	Secondary ketimine			imino-	-imine	
	Primary aldimine	RC(=NH)H		imino-	-imine	 Ethanimine
	Secondary aldimine	RC(=NR')H		imino-	-imine	
Imide	Imide	$(RCO)_2NR'$		imido-	-imide	 Succinimide (Pyrrolidine-2,5-dione)
Azide	Azide	RN_3		azido-	alkyl azide	 Phenyl azide (Azidobenzene)

Azo compound	Azo (Diimide)	RN_2R'		azo-	-diazene	 Methyl orange (p-dimethylamino-azo-benzenesulfonic acid)
Cyanates	Cyanate	ROCN		cyanato-	alkyl cyanate	 Methyl cyanate
	Isocyanate	RNCO		isocyana-to-	alkyl isocya-nate	 Methyl isocyanate
Nitrate	Nitrate	$RONO_2$		nitrooxy-, nitroxy-	alkyl **nitrate**	 Amyl nitrate (1-nitrooxypentane)
Nitrile	Nitrile	RCN		cyano-	alkanenitrile alkyl cyanide	 Benzonitrile (Phenyl cyanide)
	Isonitrile	RNC		isocyano-	alkaneisoni-trile alkyl isocya-nide	Methyl isocyanide
Nitrite	Nitrosooxy	RONO		nitroso-oxy-	alkyl **nitrite**	 Isoamyl nitrite (3-methyl-1-nitrosooxy-butane)
Nitro compound	Nitro	RNO_2		nitro-		 Nitromethane
Nitroso compound	Nitroso	RNO		nitroso-(Nitrosyl-)		 Nitrosobenzene
Oxime	Oxime	RCH=NOH	 aldoxime ketoxime		Oxime	 Acetone oxime (2-Propanone oxime)

Pyridine derivative	Pyridyl	RC$_5$H$_4$N		4-pyridyl (pyridin-4-yl)	-pyridine	
				3-pyridyl (pyridin-3-yl)		
				2-pyridyl (pyridin-2-yl)		Nicotine
Carbamate ester	Carbamate	RO(C=O)NR$_2$		(-carbamoyl)oxy-	-carbamate	Chlorpropham *(Isopropyl (3-chloro-phenyl)carbamate)*

Groups Containing Sulfur

Compounds that contain sulfur exhibit unique chemistry due to their ability to form more bonds than oxygen, their lighter analogue on the periodic table. Substitutive nomenclature (marked as prefix in table) is preferred over functional class nomenclature (marked as suffix in table) for sulfides, disulfides, sulfoxides and sulfones.

Chemical class	Group	Formula	Structural Formula	Prefix	Suffix	Example
Thiol	Sulfhydryl	RSH		sulfanyl- (-SH)	-thiol	Ethanethiol
Sulfide (Thioether)	Sulfide	RSR'		*substituent* sulfanyl- (-SR')	di(*substituent*) sulfide	(Methylsulfanyl)methane (prefix) or Dimethyl sulfide (suffix)
Disulfide	Disulfide	RSSR'		*substituent* disulfanyl- (-SSR')	di(*substituent*) disulfide	(Methyldisulfanyl)methane (prefix) or Dimethyl disulfide (suffix)
Sulfoxide	Sulfinyl	RSOR'		-sulfinyl- (-SOR')	di(*substituent*) sulfoxide	(Methanesulfinyl)methane (prefix) or Dimethyl sulfoxide (suffix)

Sulfone	Sulfonyl	RSO$_2$R'		-sulfonyl- (-SO$_2$R')	di(*substituent*) sulfone	 (Methanesulfonyl)methane (prefix) or Dimethyl sulfone (suffix)
Sulfinic acid	Sulfino	RSO$_2$H		sulfino- (-SO$_2$H)	-sulfinic acid	 2-Aminoethanesulfinic acid
Sulfonic acid	Sulfo	RSO$_3$H		sulfo- (-SO$_3$H)	-sulfonic acid	 Benzenesulfonic acid
Sulfonate ester	Sulfo	RSO$_3$R'		(-sulfonyl)oxy- or alkoxysulfonyl-	*R' R*-sulfonate	 Methyl trifluoromethanesulfonate or Methoxysulfonyl trifluoromethane (prefix)
Thiocyanate	Thiocyanate	RSCN		thiocyanato- (-SCN)	*substituent* thiocyanate	 Phenyl thiocyanate
	Isothiocyanate	RNCS		isothiocyanato- (-NCS)	*substituent* isothiocyanate	 Allyl isothiocyanate
Thioketone	Carbonothioyl	RCSR'		-thioyl- (-CSR') or sulfanylidene- (=S)	-thione	 Diphenylmethanethione (Thiobenzophenone)
Thial	Carbonothioyl	RCSH		methanethioyl- (-CSH) or sulfanylidene- (=S)	-thial	
Thiocarboxylic acid	Carbothioic *S*-acid	RC=OSH		mercaptocarbonyl-	-thioic *S*-acid	 Thiobenzoic acid (*benzothioic S-acid*)
	Carbothioic *O*-acid	RC=SOH		hydroxy(thiocarbonyl)-	-thioic *O*-acid	

			Structural Formula			Example
Thioester	Thiolester	RC=OSR'			S-alkyl-al-kane-thioate	 S-methyl thioacrylate (S-*methyl prop-2-eneth-ioate*)
	Thion-oester	RC=SOR'			O-alkyl-al-kane-thioate	
Dithiocarbox-ylic acid	Carbodith-ioic acid	RCS₂H		dithiocarboxy-	-dithioic acid	 Dithiobenzoic acid (*Benzenecarbodithioic acid*)
Dithiocarbox-ylic acid ester	Carbodithio	RC=SSR'			-dithioate	

(The RCS₂H formula uses subscript 2)

Groups Containing Phosphorus

Compounds that contain phosphorus exhibit unique chemistry due to their ability to form more bonds than nitrogen, their lighter analogues on the periodic table.

Chemical class	Group	Formula	Structural Formula	Prefix	Suffix	Example
Phosphine (Phosphane)	Phosphino	R_3P		phosphanyl-	-phosphane	 Methylpropylphosphane
Phosphonic acid	Phosphono			phosphono-	*substituent* phos-phonic acid	 Benzylphosphonic acid
Phosphate	Phosphate			phosphonooxy- or O-phosphono- (phospho-)	*substituent* phos-phate	 Glyceraldehyde 3-phos-phate (suffix) O-Phosphonocholine (prefix) (Phosphocholine)
Phosphodiester	Phosphate	HOPO(OR)₂		[(alkoxy)hydroxy-phosphoryl]oxy- or O-[(alkoxy)hydroxy-phosphoryl]-	di(*substituent*) hy-drogen phosphate or phosphoric acid di(-*substituent*) ester	DNA O[(2Guanidinoethoxy) hydroxyphosphoryl] ʟserine (prefix) (Lombricine)

Groups Containing Boron

Compounds containing boron exhibit unique chemistry due to their having partially filled octets and therefore acting as Lewis acids.

Chem-ical class	Group	Formula	Structural Formula	Prefix	Suffix	Example
Boronic acid	Borono	$RB(OH)_2$		Borono-	*substituent* boronic acid	Phenylboronic acid
Boronic ester	Boro-nate	$RB(OR)_2$		O-[bis(alkoxy)alkylbo-ronyl]-	*substituent* boronic acid di(*substituent*) ester	
Borinic acid	Borino	R_2BOH		Hydroxyborino-	di(*substituent*) borinic acid	
Borinic ester	Bori-nate	R_2BOR		O-[alkoxydialkylboro-nyl]-	di(*substituent*) borinic acid *substituent* ester	Diphenylborinic acid 2-aminoethyl ester (2-Aminoethoxydiphe-nyl borate)

Groups Containing Metals

Chemical Class	Structural Formula	Prefix	Suffix	Example
Alkyllithium	RLi		-lithium	$H_3C—Li$ methyllithium
Alkylmagnesium halide	RMgX (X=Cl, Br, I) [note 1]	(tri/di)alkyl-	-magnesium halide	$H_3C—MgCl$ methylmagnesium chloride
Alkylaluminium	Al_2R_6		-aluminium	trimethylaluminium
Silyl ether	R_3SiOR		-silyl ether	trimethylsilyl triflate

Names of Radicals or Moieties

These names are used to refer to the moieties themselves or to radical species, and also to form the names of halides and substituents in larger molecules.

When the parent hydrocarbon is unsaturated, the suffix ("-yl", "-ylidene", or "-ylidyne") replaces "-ane" (e.g. "ethane" becomes "ethyl"); otherwise, the suffix replaces only the final "-e" (e.g. "ethyne" becomes "ethynyl").

Note that when used to refer to moieties, multiple single bonds differ from a single multiple bond. For example, a methylene bridge (methanediyl) has two single bonds, whereas a methylene group (methylidene) has one double bond. Suffixes can be combined, as in methylidyne (triple bond) vs. methylylidene (single bond and double bond) vs. methanetriyl (three double bonds).

There are some retained names, such as methylene for methanediyl, 1,x-phenylene for phenyl-1,x-diyl (where x is 2, 3, or 4), carbyne for methylidyne, and trityl for triphenylmethyl.

Chemical class	Group	Formula	Structural Formula	Pre-fix	Suffix	Example
Single bond		R•		Ylo-	-yl	Methyl group Methyl radical
Double bond		R:			-ylidene	Methylidene
Triple bond		R:			-ylidyne	Methylidyne
Carboxylic acyl radical	Acyl	R–C(=O)•			-oyl	Acetyl

Aliphatic Compound

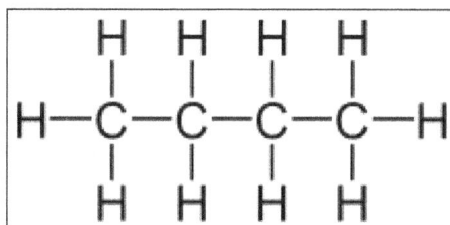

Acyclic aliphatic/non-aromatic compound (butane).

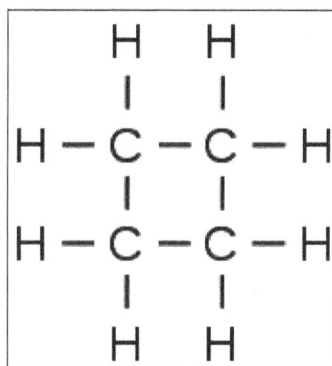

Cyclic aliphatic/non-aromatic compound (cyclobutane).

In organic chemistry, hydrocarbons (compounds composed of carbon and hydrogen) are divided into two classes: aromatic compounds and aliphatic compounds, also known as non-aromatic compounds. Aliphatics can be cyclic, but if a unique type of especially stable cyclic bond exists in the molecule, called a benzene ring, then it is considered to be an aromatic compound. Aliphatic compounds can be saturated, like hexane, or unsaturated, like hexene and hexyne. Open-chain compounds (whether straight or branched) contain no rings of any type, and are thus aliphatic.

Structure

Aliphatic compounds can be saturated, joined by single bonds (alkanes), or unsaturated, with double bonds (alkenes) or triple bonds (alkynes). Besides hydrogen, other elements can be bound to the carbon chain, the most common being oxygen, nitrogen, sulfur, and chlorine.

The least complex aliphatic compound is methane (CH_4).

Properties

Most aliphatic compounds are flammable, allowing the use of hydrocarbons as fuel, such as methane in Bunsen burners and as liquefied natural gas (LNG), and ethyne (acetylene) in welding.

Examples of Aliphatic Compounds/Non-Aromatic

The most important aliphatic compounds are:

- n-, iso- and cyclo-alkanes (saturated hydrocarbons).

- n-, iso- and cyclo-alkenes and -alkynes (unsaturated hydrocarbons).

Important examples of low-molecular aliphatic compounds can be found in the list below (sorted by the number of carbon-atoms):

Formula	Name	Structural Formula	Chemical Classification
CH_4	Methane		Alkane
C_2H_2	Acetylene	H—C≡C—H	Alkyne
C_2H_4	Ethylene		Alkene
C_2H_6	Ethane		Alkane
C_3H_4	Propyne		Alkyne

C_3H_6	Propene		Alkene
C_3H_8	Propane		Alkane
C_4H_6	1,2-Butadiene		Diene
C_4H_6	1-Butyne		Alkyne
C_4H_8	1-Butene		Alkene
C_4H_{10}	Butane		Alkane
C_6H_{10}	Cyclohexene		Cycloalkene
C_5H_{12}	*n*-pentane		Alkane
C_7H_{14}	Cycloheptane		Cycloalkane
C_7H_{14}	Methylcyclohexane		Cyclohexane
C_8H_8	Cubane		Octane
C_9H_{20}	Nonane		Alkane
$C_{10}H_{12}$	Dicyclopentadiene		Diene, Cycloalkene
$C_{10}H_{16}$	Phellandrene		Terpene, Diene Cycloalkene

$C_{10}H_{16}$	α-Terpinene		Terpene, Cycloalkene, Diene
$C_{10}H_{16}$	Limonene		Terpene, Diene, Cycloalkene
$C_{11}H_{24}$	Undecane		Alkane
$C_{30}H_{50}$	Squalene		Terpene, Polyene
$C_{2n}H_{4n}$	Polyethylene		Alkane

Organic Reaction

Organic reactions are chemical reactions involving organic compounds. The basic organic chemistry reaction types are addition reactions, elimination reactions, substitution reactions, pericyclic reactions, rearrangement reactions, photochemical reactions and redox reactions. In organic synthesis, organic reactions are used in the construction of new organic molecules. The production of many man-made chemicals such as drugs, plastics, food additives, fabrics depend on organic reactions.

The oldest organic reactions are combustion of organic fuels and saponification of fats to make soap. Modern organic chemistry starts with the Wöhler synthesis in 1828. In the history of the Nobel Prize in Chemistry awards have been given for the invention of specific organic reactions such as the Grignard reaction in 1912, the Diels-Alder reaction in 1950, the Wittig reaction in 1979 and olefin metathesis in 2005.

Classifications

Organic chemistry has a strong tradition of naming a specific reaction to its inventor or inventors and a long list of so-called named reactions exists, conservatively estimated at 1000. A very old named reaction is the Claisen rearrangement (1912) and a recent named reaction is the Bingel reaction (1993). When the named reaction is difficult to pronounce or very long as in the Corey-House-Posner-Whitesides reaction it helps to use the abbreviation as in the CBS reduction. The number of reactions hinting at the actual process taking place is much smaller, for example the ene reaction or aldol reaction.

Another approach to organic reactions is by type of organic reagent, many of them inorganic, required in a specific transformation. The major types are oxidizing agents such as osmium tetroxide, reducing agents such as Lithium aluminium hydride, bases such as lithium diisopropylamide and acids such as sulfuric acid.

Finally, reactions are also classified by mechanistic class. Commonly these classes are (1) polar, (2) radical, and (3) pericyclic. Polar reactions are characterized by the movement of electron pairs from a well-defined source (a nucleophilic bond or lone pair) to a well-defined sink (an electrophilic center with a low-lying antibonding orbital). Participating atoms undergo changes in charge, both in the formal sense as well as in terms of the actual electron density. The vast majority of organic reactions fall under this category. Radical reactions are characterized by species with unpaired electrons (radicals) and the movement of single electrons. Radical reactions are further divided into chain and nonchain processes. Finally, pericyclic reactions involve the redistribution of chemical bonds along a cyclic transition state. Although electron pairs are formally involved, they move around in a cycle without a true source or sink. These reactions require the continuous overlap of participating orbitals and are governed by orbital symmetry considerations. Of course, some chemical processes may involve steps from two (or even all three) of these categories, so this classification scheme is not necessarily straightforward or clear in all cases. Beyond these classes, transition-metal mediated reactions are often considered to form a fourth category of reactions, although this category encompasses a broad range of elementary organometallic processes, many of which have little in common.

Fundamentals

Factors governing organic reactions are essentially the same as that of any chemical reaction. Factors specific to organic reactions are those that determine the stability of reactants and products such as conjugation, hyperconjugation and aromaticity and the presence and stability of reactive intermediates such as free radicals, carbocations and carbanions.

An organic compound may consist of many isomers. Selectivity in terms of regioselectivity, diastereoselectivity and enantioselectivity is therefore an important criterion for many organic reactions. The stereochemistry of pericyclic reactions is governed by the Woodward–Hoffmann rules and that of many elimination reactions by the Zaitsev's rule.

Organic reactions are important in the production of pharmaceuticals. In a 2006 review it was estimated that 20% of chemical conversions involved alkylations on nitrogen and oxygen atoms, another 20% involved placement and removal of protective groups, 11% involved formation of new carbon-carbon bond and 10% involved functional group interconversions.

By Mechanism

There is no limit to the number of possible organic reactions and mechanisms. However, certain general patterns are observed that can be used to describe many common or useful reactions. Each reaction has a stepwise reaction mechanism that explains how it happens, although this detailed description of steps is not always clear from a list of reactants alone. Organic reactions can be organized into several basic types. Some reactions fit into more than one category. For example, some substitution reactions follow an addition-elimination pathway. This overview isn't intended to include every single organic reaction. Rather, it is intended to cover the basic reactions.

Reaction type	Subtype	Comment
Addition reactions	electrophilic addition	include such reactions as halogenation, hydrohalogenation and hydration.
	nucleophilic addition	
	radical addition	
Elimination reaction		include processes such as dehydration and are found to follow an E1, E2 or E1cB reaction mechanism
Substitution reactions	nucleophilic aliphatic substitution	with S_N1, S_N2 and S_Ni reaction mechanisms
	nucleophilic aromatic substitution	
	nucleophilic acyl substitution	
	electrophilic substitution	
	electrophilic aromatic substitution	
	radical substitution	
Organic redox reactions		are redox reactions specific to organic compounds and are very common.
Rearrangement reactions	1,2-rearrangements	
	pericyclic reactions	
	metathesis	

In condensation reactions a small molecule, usually water, is split off when two reactants combine in a chemical reaction. The opposite reaction, when water is consumed in a reaction, is called hydrolysis. Many polymerization reactions are derived from organic reactions. They are divided into addition polymerizations and step-growth polymerizations.

In general the stepwise progression of reaction mechanisms can be represented using arrow pushing techniques in which curved arrows are used to track the movement of electrons as starting materials transition to intermediates and products.

Other Classification

In heterocyclic chemistry, organic reactions are classified by the type of heterocycle formed with respect to ring-size and type of heteroatom. Reactions are also categorized by the change in the carbon framework. Examples are ring expansion and ring contraction, homologation reactions, polymerization reactions, insertion reactions, ring-opening reactions and ring-closing reactions.

Organic reactions can also be classified by the type of bond to carbon with respect to the element involved. More reactions are found in organosilicon chemistry, organosulfur chemistry, organophosphorus chemistry and organofluorine chemistry. With the introduction of carbon-metal bonds the field crosses over to organometallic chemistry.

References

- Brown, Theodore (2002). Chemistry: the central science. Upper Saddle River, NJ: Prentice Hall. P. 1001. ISBN 0130669970

- What-is-an-organic-compound-13712143: sciencing.com, Retrieved 1 April, 2019

- Strategic Applications of Named Reactions in Organic Synthesis Laszlo Kurti, Barbara Czako Academic Press (March 4, 2005) ISBN 0-12-429785-4

- Analysis of the reactions used for the preparation of drug candidate moleculesjohn S. Carey, David Laffan, Colin Thomson and Mike T. Williams Org. Biomol. Chem., 2006, 4, 2337–2347, doi:10.1039/b602413k

4

Inorganic Chemistry

The chemical compounds that lack carbon-hydrogen bonds are known as inorganic compounds. Inorganic chemistry focuses on the behavior and synthesis of inorganic and organometallic compounds. The topics elaborated in this chapter will help in gaining a better perspective about inorganic chemistry.

The word organic refers to the compounds which contain the carbon atoms in it. So the branch of chemistry that deals with the study of compounds, which does not consist of carbon-hydrogen atoms in it, is called as 'Inorganic Chemistry.' In simple words, it is opposite to that of the Organic Chemistry. The substances which do not have carbon-hydrogen bonding are the metals, salts, chemical substances, etc.

On this planet, there are known to exist about 100,000 number of Inorganic compounds. Inorganic chemistry studies the behavior of these compounds along with their properties, their physical and chemical characteristics too. The elements of the periodic table except for carbon and hydrogen, come in the lists of Inorganic compounds.

Classification of Inorganic Compounds

The organic compounds that are classified under Inorganic chemistry are:

- Acids: Acids are those compounds that dissolve in water and generate hydrogen ions or H^+ Ions. The examples of acids include Hydrochloric acid, citric acid, sulphuric acid, vinegar, etc. One example of the acidic reaction is shown below:

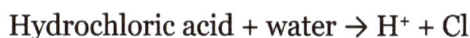

  Hydrochloric acid + water $\rightarrow H^+ + Cl$

- Bases: A base is a type of substance or a compound that produces hydroxyl ions when kept in water. The bases like potassium hydroxide, calcium hydroxide, ammonia, sodium hydroxide produce OH^- ions when dissolved in water.

  Potassium Hydroxide + $H_2O \rightarrow K^+ + OH^-$

- Salts: As you might be familiar with the word 'Salt'. The substances obtained as a result of the reaction between an acid and a base are called as Salts. The table salt of the sodium hydroxide is one of the typical examples of salts.

- Oxides: The compounds which consist of one oxygen atom called as Oxides.

Types of Reactions and Examples

There are about four types chemical reactions of Inorganic chemistry namely combination, decomposition, single displacement and double displacement reactions.

- Combination Reactions: As it is in the name 'Combination', here two or more substances combine to form a product which is called as Combination reaction. For example:

  Barium $+ F_2 \rightarrow BaF_2$

- Decomposition Reaction: It is a type of reaction where a single element splits up or say decomposes into two products. For example:

  $FeS \rightarrow Fe + S$

- Single Displacement Reactions: A reaction where a single atom of one element replaces another atom of one more element. For example:

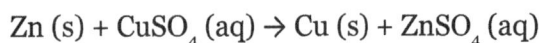

  $Zn \ (s) + CuSO_4 \ (aq) \rightarrow Cu \ (s) + ZnSO_4 \ (aq)$

- Double Displacement Reactions: This type of reaction is also called as 'metathesis reactions'. Here two elements of two different compounds displace each other to form two new compounds. For example:

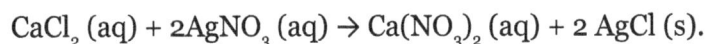

  $CaCl_2 \ (aq) + 2AgNO_3 \ (aq) \rightarrow Ca(NO_3)_2 \ (aq) + 2 \ AgCl \ (s)$.

Applications of Inorganic Chemistry

Inorganic chemistry finds its high number of applications in various fields such as Biology, chemical, engineering, etc.

- It is applied in the field of medicine and also in healthcare facilities.

- The most common application is the use of common salt or the compound Sodium hydroxide in our daily lives.

- Baking soda is used in the preparation of cakes and other foodstuffs.

- Many inorganic compounds are utilized in ceramic industries.

- In the electrical field, it is applied to the electric circuits as silicon in the computers, etc.

Nomenclature of Inorganic Chemistry

The purpose of nomenclature in chemistry is to convey information about the material being described. The designation chosen should be unequivocal, at least within the limitations of the type of nomenclature adopted. The type adopted will depend in part on the total amount of information to be conveyed, the kind of compound to be described, and the whim of the person describing the compound.

Nomenclaturists use the terms "trivial" and "systematic" to describe two major divisions of nomenclature. Systematic nomenclature is based on established principles so that it can be extended in a logical way to describe known, new, and hypothetical compounds. A trivial nomenclature is one established by rule of thumb and includes many of the older names (spirit of salt, aqua regia, etc.) and lab nomenclatures (the green chelate, etc.).

Actual usage is often a mixture of the two types, and the fundamental bases of all chemical names, those of the elements, are essentially trivial. Note that the name methane is trivial, but that the name pentane is not. For this discussion, a formula representing a compound can be regarded simply as a kind of name. The principal general (but by no means the only) types of nomenclature used in inorganic chemistry are substitutive and additive (coordination).

Substitutive Nomenclature

Substitutive nomenclature is essentially an organic invention and follows the historical development of organic chemistry. It starts with the designation of an appropriate parent compound from which the compound under discussion can be developed formally by substitution or replacement processes. In organic chemistry these parents can be the paraffins, and in inorganic chemistry they are generally (and arbitrarily) taken to be the hydrides of the elements of Periodic Groups 14, 15, and 16, plus boron, which also has an additional rather specific nomenclature of its own. Thus the formula SiH_3Cl can be named chlorosilane, as a substituted derivative of the saturated parent SiH_4, silane (compare chloromethane). The generation of a radical by the loss of a hydrogen atom from the parent is indicated by modification of the termination, silane becoming silyl, $SiH_3 \cdot$ (the superscript dot indicates an unpaired electron). The name silyl can be used to represent a substituent group in another parent hydride (compare methyl) or for the unbound radical, and the procedure is quite general for all parent hydrides to which the methodology is applied. Silane can also be modified formally by the removal of a proton, yielding the anion SiH_3^-. The name then takes the characteristic anion ending -ide: silanide. The formal addition of a proton is indicated by another termination (-ium), giving SiH_5^+, silanium. These terminations are used generally in inorganic nomenclature, as in chloride for Cl^- and ammonium for NH_4^+. Other formal operations recognized in substitutive nomenclature include addition or removal of a hydride from the parent. This can be indicated by the termination -ylium, giving the name silylium for SiH_3^+.

Other Modifications of Names

The terminations cited above can be used generally in inorganic nomenclature. However, they are sometimes not applicable, especially where parent hydrides are not reasonably definable. Inorganic chemists have tended to assign electropositive and electronegative character to elements, though numerical values are not necessarily easy to define. Metals are generally assigned electropositive character and nonmetals electronegative character. The names developed on this basis may imply formally a saltlike nature even in compounds that are not really salts at all. Thus common salt is called sodium chloride, which is ionic, but phosphorus trichloride is certainly not saltlike. It is not wise to infer the detailed physical nature of a compound from the name alone. In this system the name of the (electropositive) metal is not modified from that of the element, but the name of the electronegative element is, and in the way described in the Substitutive Nomenclature section, above. Similarly we derive oxide and sulfide, for example, from oxygen and sulfur. The same division between electronegative and electropositive parts is evident in the covalent nonionic compound $SiCl_4$, which can be named silicon tetrachloride , though an equally valid substitutive name is tetrachlorosilane.

Inorganic chemists also use a further termination to indicate the name of a cation. This is the ending -ate, and it is used as a modification of the name of an oxoacid. Thus sulfuric acid, H_2SO_4, gives rise to sulfate, SO_4^{2-}, phosphoric acid to phosphate, PO_4^{3-}, and nitric acid to nitrate, NO_3^-. The partially deprotonated anions such as HSO_4^- and $H_2PO_4^{2-}$ are rather more complicated to deal with, and are discussed in Nomenclature of Inorganic Chemistry, often referred to as the Red Book.

In an older procedure that is no longer recommended, the name of an electropositive element displaying more than one oxidation state in its compounds was sometimes modified to indicate the particular oxidation state involved. Thus iron chlorides were often named ferrous chloride and ferric chloride to convey the two oxidation states of II and III (note that, like normal arabic numbers, these Roman numerals are positive unless otherwise shown by a negative sign). However, the use was not consistent. Cuprous and cupric chlorides indicated oxidation states I and II, and phosphorous and phosphoric chlorides indicated oxidation states III and V. Modern nomenclature specifies the oxidation state of the electropositive partner in these compounds directly: iron(II) chloride, iron(III) chloride, copper(I) chloride, copper(II) chloride, phosphorus(III) chloride, and phosphorus(V) chloride. These designations are unequivocal. The number of counter anions, 1, 2, 3, or 5, should immediately be evident. Examples of negative oxidation states include oxide(−II) or oxide(2−), and dioxide(−I) or dioxide(1−). Note that in a multi-atom group, of which PO_4^{3-} may be taken as an example, the charge on any given atom may not be evident, even if the overall charge is known. In contrast, the oxidation states phosphorus(V) and oxide(−II) are much more readily defined. The use of such charges in names and formulae in these circumstances is not recommended.

Formulae

The rules for formulae for the compounds discussed above are rather elastic. At its simplest, a formula is a list of element symbols accompanied by multiplying subscripts indicating the atomic proportions of each kind of atom. These formulae may be empirical, simply corresponding to the atom ratios, or stoichiometric, representing the totality of the atoms within a molecule. The latter can be used to calculate a molecular weight. Strictly speaking, for a compound that exists as discrete molecules, this latter can also be termed a molecular formula, but this is a misnomer for ionic compounds and for compounds of which the structure changes with temperature. The ordering of these symbols can be adjusted to suit the requirements of the user. At the simplest, an alphabetical order is used, since this is the same in most European languages. Many chemists emphasize the importance of carbon and hydrogen and adopt a sequence C, H, N, and then the remaining element symbols in alphabetical order. Such devices are often employed in indexes. Inorganic chemists often group the atomic symbols in a formula in electropositive and electronegative groups, designated as discussed above. This can be a somewhat arbitrary procedure, and the relative positions of atoms in an electronegativity sequence may be established using the Periodic table. For simple cases, formulae such as NaCl or $SiCl_4$ are used. Anionic groups are assumed to be electronegative, hence $Ca_3(PO_4)_2$. The parentheses are used to define the associated groups of atoms within the formula.

Formulae can also be used to indicate two- or three-dimensional structures. This is particularly useful for coordination compounds, which are discussed next. However, this use is not restricted to classical coordination compounds, as the following examples show. Special devices are often adopted to indicate bonds or lines that are not in the plane of the paper. Their use is not consistent throughout chemistry, but the meaning in any given case is generally obvious.

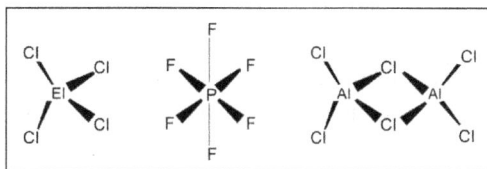

The first example represents a tetrahedral arrangement, because the solid defined by the four chlorine atoms at its apices is a tetrahedron. The second is octahedral , and the third represents two edge-fused tetrahedra. The wedge bonds are pointing in front of or behind the plane of the paper; the thin lines designate bonds in the plane of the paper.

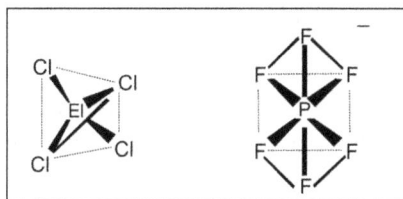

Inorganic chemists often represent tetrahedra, octahedra, and other shapes in their formulae, to help the reader identify molecular shapes. The broken lines designating these shapes are not intended to represent bonds between atoms.

Oxidation states may also be indicated in formulae where this is helpful, though the need to do so is not common in the simplest cases. The following examples show the formalism employed: $Fe^{II}Cl_2$, $Fe^{III}Cl_3$, $Cu^{I}Cl$, $Cu^{II}Cl_2$, $P^{III}Cl_3$, $P^{V}Cl_5$.

Coordination Nomenclature

This is an additive nomenclature, and just as organic chemists have developed substitutive nomenclature in parallel with the methodology of substitutive chemistry, inorganic chemists have developed a nomenclature for coordination compounds that arises from the formal assembly of a coordination entity from its components, a central metal ion (in the simplest cases) and its ligands. Such a coordination entity may be neutral or it may carry a charge, positive or negative. Any such charge may be shown in the usual way, using formalisms such as 2– and 3–. Clearly organometallic compounds , depending upon their type, may be named either from substituted parent hydrides or as coordination entities.

Formulae in Coordination Nomenclature

The general rule is that the formula of a coordination entity should always appear within square brackets, even when the entity itself is an infinite polymer. The use of enclosing marks (square brackets, curly brackets, and parentheses) is slightly different for the usage that is common in organic chemistry. The usual priority sequence is [( )], [{( )}], [{[( )]}], [{{[( )]}}], and so on. Brackets should always be used if they make the formula clearer. The order of symbol citation within the formula of a coordination entity should begin with the metal ion followed by the ligands, ideally with charged ligands cited in alphabetical order using the first symbol of the ligand formula, and these are then followed as a class by the neutral ligand formulae, similarly ordered. The division into neutral and charged ligands can be somewhat arbitrary. Since a ligand is generally assumed to present a lone pair of electrons to the central metal, groups such as CH_3 are formally regarded as anions rather than as radicals with unpaired electrons, even though they usually carry the names of radicals. Compounds that really do possess unpaired electrons in the free state can cause problems, especially when calculating oxidation states. For coordination nomenclature purposes, NO, nitrogen(II) oxide, is considered to be a neutral ligand. Complicated ligands may be represented by abbreviations rather than formulae, and lists of recommended abbreviations have been published in sources such as *Nomenclature of Inorganic Chemistry*. Some examples of these usages are shown in table. The use of square brackets to indicate the coordination entity is fundamental and is a particularly useful device.

Note the negative oxidation state and the η (hapto) connectivity symbol in the last two examples. Where appropriate, stereochemical descriptors, such as *cis -*, *trans -*, *mer -*, and *fac -*, polyhedral descriptors, and chirality descriptors may be added to give structural information, but these are more often used in names, except for the simplest formulae. Polynuclear species may be described using the appropriate multiplicative suffixes, and bridging ligands can also be shown. The bridging symbol μ_n is useful for this purpose.

Compound formulae	Complex ion formulae	Showing oxidation state
$[Co(NH_3)_6]Cl_3$	$[Co(NH_3)_6]^{3+}$	$[Co^{III}(NH_3)_6]^{3+}$
$[CoCl(NH_3)_5]Cl_2$	$[CoCl(NH_3)_5]^{2+}$	$[Co^{III}Cl(NH_3)_5]^{2+}$
$[CoCl(NO_2)(NH_3)_4]Cl$	$[CoCl(NO_2)(NH_3)_4]^{+}$	$[Co^{III}Cl(NO_2)(NH_3)_4]^{+}$
$[PtCl(NH_2CH_3)(NH_3)_2]Cl$	$[PtCl(NH_2CH_3)(NH_3)_2]^{+}$	$[Pt^{II}Cl(NH_2CH_3)(NH_3)_2]^{+}$
$[CuCl_2\{O=C(NH_2)_2\}_2]$		$[Cu^{II}Cl_2\{O=C(NH_2)_2\}_2]$
$K_2[PdCl_4]$	$[PdCl_4]^{2-}$	$[Pd^{II}Cl_4]^{2-}$
$K_2[OsCl_5N]$	$[OsCl_5N]^{2-}$	$[Os^{VI}Cl_5N]^{2-}$
$Na[PtBrCl(NO_2)(NH_3)]$	$[PtBrCl(NO_2)(NH_3)]^{-}$	$[Pt^{II}BrCl(NO_2)(NH_3)]^{-}$
$[Co(en)_3]Cl_3$	$[Co(en)_3]^{3+}$	$[Co^{III}(en)_3]^{3+}$
$Na_2[Fe(CO)_4]$	$[Fe(CO)_4]^{2-}$	$[Fe^{-II}(CO)_4]^{2-}$
$[Co(\eta^5-C_5H_5)_2]Cl$	$[Co(\eta^5-C_5H_5)_2]^{+}$	$[Co^{II}(\eta^5-C_5H_5)_2]^{+}$

The subscript may be omitted if a ligand bridges only two groups. Polymeric materials can be indicated in an empirical formula using the indeterminate subscript n. When there are different central metal ions present in a polynuclear compound, the established priority sequence for metal ions should be used to determine the order of citation.

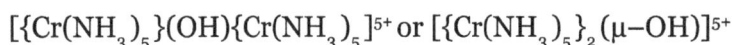

$[\{Cr(NH_3)_5\}(OH)\{Cr(NH_3)_5\}]^{5+}$ or $[\{Cr(NH_3)_5\}_2(\mu-OH)]^{5+}$

$[Re_2Br_8]^{4-}$ or $[(ReBr_4)_2]^{4-}$

$[[IrCl_2(CO)\{P(C_6H_5)_3\}_2](HgCl)]$

$[\{PdCl_2\}_n]$ or $[\{Pd(\mu-Cl)_2\}_n]$.

Names in Coordination Nomenclature

The names of coordination entities are assembled using principles similar, but not identical, to those used for formulae. The central atom is always cited last. Its name may be modified by an oxidation state symbol. The ligands are presented in the alphabetical order of their initial letters, neglecting for this purpose any multiplicative prefixes. It is not necessary to divide the ligands into neutral and charged groups. However, the names of negatively charged ligands are generally modified by adding the postfix suffix -o in place of the final -e where it occurs, to indicate that they are indeed bound and not free. As an exception, this is not the case with hydrocarbon ligands such as methyl and ethyl, which retain the names of radicals. The names of neutral ligands are not modified. If the coordination entity itself is negatively charged (but not when it is neutral or positively charged), then the name of the central atom is modified by the ending -ate. These practices are illustrated below:

$[Co(NH_3)_6]Cl_3$	hexaamminecobalt(III) trichloride
$[Co(NH_3)_6]^{3+}$	hexaamminecobalt(3+)
$[CoCl(NH_3)_5]Cl_2$	pentaamminechlorocobalt(III) trichloride
$[CoCl(NH_3)_5]^{2+}$	pentaamminechlorocobalt(2+)
$[CoCl(NO_2)(NH_3)_4]Cl$	tetraamminechloronitritocobalt(III) chloride
$[CoCl(NO_2)(NH_3)_4]^+$	tetraamminechloronitritocobalt(1+)
$[PtCl(NH_2CH_3)(NH_3)_2]Cl$	bisamminechloromethylamineplatinum(II) chloride
$[PtCl(NH_2CH_3)(NH_3)_2]^+$	diamminechloromethylamineplatinum(+)
$[CuCl_2\{O=C(NH_2)_2\}_2]$	dichlorobis(urea)copper(II)
$K_2[PdCl_4]$	potassium tetrachloropalladate(II)
$K_2[OsCl_5N]$	potassium pentachloronitrodoosmate(VI)
$[Co(H_2O)_2(NH_3)_4]Cl$	tetraamminediaquacobalt(III) chloride

Note that in some cases it may be useful to introduce additional enclosing marks to ensure clarity: for example, to avoid possible confusion between chloromethylamine, $ClCH_2NH_2$, and (chloro)methylamine, which implies two separate ligands, Cl and CH_3NH_2. It is for the writer to decide whether such a strategy is useful, depending on the particular case under review. Ammonia as a ligand has the name ammine. Similarly, water has the coordination name aqua.

$Na[PtBrCl(NO_2)(NH_3)]$	sodium amminebromochloronitrito-platinate(II)
$[Co(en)_3]Cl_3$	tris(ethane-1,2-diamine)cobalt(II) trichloride
$Na_2[Fe(CO)_4]$	sodium tetracarbonylferrate(−II)
$[Co(\eta^5-C_5H_5)_2]Cl$	bis(cyclopentadienyl)cobalt(III) chloride or bis(η^5-cyclopentadienyl)cobalt(III) chloride.

The symbol η is used above and also quite generally throughout organometallic coordination chemistry to indicate the number of carbon atoms in a ligand that are coordinated to the metal. Other devices to indicate connectivity are the italicized atomic symbols of the donor atoms (useful for indicating structure in complexes containing chelating and polydentate ligands), and for some complicated cases, the κ symbolism may be useful.

Further devices are used in coordination names to show polymeric structures, which may contain bridging groups and metal-metal bonds.

$[\{Pd(\mu-Cl)_2\}_n]$	poly(di-μ-chloropalladium)
$[\{Cr(NH_3)_5\}(OH)\{Cr(NH_3)_5\}]^{5+}$	μ-hydroxo-bis[pentaam-minechromium(III)](5+)
$[(ReBr_4)_2]^{4-}$	bis(tetrabromorhenate)(Re-Re)(2−)
$[[IrCl_2(CO)\{P(C_6H_5)_3\}_2](HgCl)]$	carbonyl-1 κC-trichloro-1 κ^2,2 κCl-bis(triphenylphosphine-1 κP)iridium mercury(Hg-Ir)

Where different metals are present, priority rules must be applied to assign metal locants.

Inorganic Compound

An inorganic compound is any compound that lacks a carbon atom, for lack of a more in-depth definition. Those compounds with a carbon atom are called organic compounds, due to their root base in an atom that is vital for life. There are a small number of inorganic compounds that actually do contain carbon, given its propensity for forming molecular bonds; these include carbon monoxide and carbon dioxide, to name a few.

Inorganic compounds are often quite simple, as they do not form the complex molecular bonds that carbon makes possible. A common example of a simple inorganic compound would be sodium chloride, known more commonly as household salt. This compound contains only two atoms, sodium (Na) and chlorine (Cl).

Examples of Inorganic Compounds:

- H_2O: Water is a simple inorganic compound, even though it contains hydrogen, a key atom (along with carbon) in many organic compounds. The atoms in a molecule of water have formed very simple bonds due to this lack of carbon.

- HCl: Hydrochloride, also known as hydrochloric acid when it is dissolved in water, is a colorless, corrosive acid with a fairly strong pH. It is found in the gastric juices of many animals, helping in digestion by breaking down food.

- CO_2: Carbon dioxide, despite the presence of a carbon atom in the formula, is classified as an inorganic compound. This has caused a dispute within the scientific community, with questions being raised as to the validity of our current methods of classifying compounds. Currently, organic compounds contain a carbon or a hydrocarbon, which forms a stronger bond. The bond formed by carbon in CO_2 is not a strong bond.

- NO_2: Nitrogen dioxide gas presents a variety of colors at different temperatures. It is often produced in atmospheric nuclear tests, and is responsible for the tell-tale reddish color displayed in mushroom clouds. It is highly toxic, and forms fairly weak bonds between the nitrogen and oxygen atoms.

- Fe_2O_3: Iron (III) oxide is one of the three main oxides of iron, and is an inorganic compound due to the lack of a carbon atom or a hydrocarbon. Iron (III) oxide occurs naturally as hematite, and is the source of most iron for the steel production industry. It is commonly known as rust, and shares a number of characteristics with its naturally occurring counterpart.

5

Acids and Bases

Acid is a molecule that is capable of donating a proton and forming a covalent bond with a pair of electrons. Bases are the substances that release hydroxide ions in an aqueous solution. This chapter has been carefully written to provide an easy understanding of acids and bases as well as acid-base reactions.

Acid

Acids are a molecule or other species which can donate a proton or accept an electron pair in reactions. All acid elements have a few things in common i.e all are sour in taste, they turn blue litmus paper to red, and lose their acidity if they're combined with alkaline substances. The pH level of acids ranges from $0 - 6$.

Some common examples of acids are Citrus fruits such as lemons, limes, oranges, grapefruit, etc. All these fruits contain citric acid. Hence, they taste sour or tart. Citric acid is a weak acid but still, it produces hydrogen ions when mixed with water and that's why the pH of lemon juice is 2. Another example of an acid is vinegar. Vinegar consists of acetic acid. Ever wondered why your skin becomes red and swollen after an ant bite or a mosquito bite? Because these insects inject formic acid which causes such skin reactions. Nitric acid, sulfuric acid, hydrochloric acid, etc. are other common ones.

Properties of Acids

The properties of Acids are as follows:

- Acids change the color of blue litmus to red.

- They change the color of Methyl Orange/Yellow to Pink.

- Acidic substances converts Phenolphthalein from deep pink to colorless.

- Are sour or tart in taste.

- The pH level of acids range from 0-6.

- Acids lose their acidity when combined with alkalines.

- They destroy the chemical property of bases.

- When reacting with metals they produce hydrogen gas.

- Acids produce carbon dioxide when reacted with carbonates.

- Most acids are corrosive in nature which means that they tend to corrode or rust metals.

Classification of Acids

Acids are often classified on the basis of source, the presence of oxygen, strength, concentration and basicity.

Classification based on the Source

This means that the acid is classified on the basis of their source or origin. They are mainly of two types: Organic acid and Mineral acid.

- Organic Acid: This is the acid obtained from organic materials such as plants and animals. For e.g. Citric acid (Citrus fruits), Acetic acid (Vinegar), Oleic acid (Olive oil), etc.

- Mineral Acid: Mineral acid is procured from minerals. They are also known as inorganic acids. They do not contain carbon. For e.g. H_2SO_4, HCl. HNO_3, etc.

Classification based on the Presence of Oxygen

This means that the acids are classified on the basis of the presence of oxygen. These are of two types: Oxy-acid and Hydracids.

- Oxy-acid: Acids that consist oxygen in their composition is known as Oxy-acids. For e.g. H_aSO_4, HNO_3, etc.

- Hydracid: Those that consist hydrogen combined with other elements and do not contain any oxygen in their composition and do not contain any oxygen in their composition are called Hydracids. For e.g. HCl, HI, HBr, etc.

Classification based on the Strength of the Acid

Acids produce hydrogen ions when mixed with H2O, the strength of an acid depends on its concentration of the hydrogen ions present in a solution. A greater number of hydrogen ions means greater strength of the acid whereas, lower number of hydrogen ions means that the acid is weak. They are classified as:

- Strong Acids: An acid which can be dissociated completely or almost completely in water is known as a strong acid. For e.g. sulphuric acid, nitric acid, hydrochloric acid, etc.

  $H^+ + H_2O <\text{---}> H_3O^+$

  $HCl(aq) <\text{---}> H+(aq) + SO_4^-(aq)$

  $H_2SO_4(aq) <\text{---}> 2H+(aq) + SO_4^-(aq)$

- Weak Acids: An acid which doesn't dissociate completely or dissociates negligibly in water is known as a weak acid. For e.g. Those that which we usually consume on daily basis i.e. citric acid, acetic acid, etc.

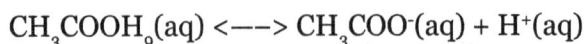

$CH_3COOH_9(aq) <--> CH_3COO^-(aq) + H^+(aq)$

$HCOOH(aq) <-> HCOO^-(aq) + H^+(aq)$

Classification based on its Concentration

As we have studied above, the concentration of the acid depends on the number of hydrogen ions that it produces in water. Based on this the acid is classified as:

- Concentrated Acid: When an aqueous solution has a relatively high percentage of acid dissolved in it, then it is a concentrated acid. For e.g. concentrated hydrochloric acid, concentrated sulphuric acid, concentrated nitric acid, etc.

- Diluted Acid: When an aqueous solution has a relatively low percentage of acid dissolved in it, then it is a dilute acid. For e.g. dilute hydrochloric acid, dilute sulphuric acid, dilute nitric acid, etc.

Classification based on the Basicity of the Acid

Acid on dissociation in water produces hydrogen ion. The number of these hydrogen ions that can be replaced in an acid is the basicity of an acid.

- Monobasic Acid: A monobasic acid is an acid which has only one hydrogen ion. Therefore, these acids combine with one hydroxyl group of the base to form salt and water. For e.g. HCl, HCOOH, HBr, etc.

- Dibasic Acid: Dibasic acid is that which shares twp hydroxyl groups it is known as dibasic acid. Dibasic acid dissociates in 2 steps. They can provide 2 kinds of salts i.e. the normal salt and a hydrogen salt.

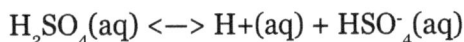

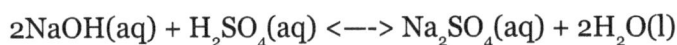

  $H_2SO_4(aq) <-> H+(aq) + HSO^-_4(aq)$

  $2NaOH(aq) + H_2SO_4(aq) <--> Na_2SO_4(aq) + 2H_2O(l)$

- Tribasic Acid: Tribasic acids are those which can combine with three hydroxyl groups. They have three replaceable hydrogen ions, and they produce 3 types of salts. For e.g. H3PO4.

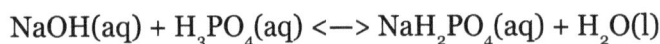

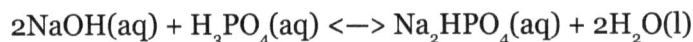

  $NaOH(aq) + H_3PO_4(aq) <-> NaH_2PO_4(aq) + H_2O(l)$

  $2NaOH(aq) + H_3PO_4(aq) <-> Na_2HPO_4(aq) + 2H_2O(l)$

Bases

In chemistry, bases are substances that, in aqueous solution, release hydroxide (OH^-) ions, are slippery to the touch, can taste bitter if an alkali, change the color of indicators (e.g., turn red litmus paper blue), react with acids to form salts, promote certain chemical reactions (base catalysis), accept protons from any proton donor or contain completely or partially displaceable OH⁻ ions. Examples of bases are the hydroxides of the alkali metals and the alkaline earth metals

In water, by altering the autoionization equilibrium, bases yield solutions in which the hydrogen ion activity is lower than it is in pure water, i.e., the water has a pH higher than 7.0 at standard conditions. A soluble base is called an alkali if it contains and releases OH⁻ ions quantitatively. However, it is important to realize that basicity is not the same as alkalinity. Metal oxides, hydroxides, and especially alkoxides are basic, and conjugate bases of weak acids are weak bases.

Bases can be thought of as the chemical opposite of acids. However, some strong acids are able to act as bases. Bases and acids are seen as opposites because the effect of an acid is to increase the hydronium (H_3O^+) concentration in water, whereas bases reduce this concentration. A reaction between an acid and a base is called neutralization. In a neutralization reaction, an aqueous solution of a base reacts with an aqueous solution of an acid to produce a solution of water and salt in which the salt separates into its component ions. If the aqueous solution is saturated with a given salt solute, any additional such salt precipitates out of the solution.

For a substance to be classified as an Arrhenius base, it must produce hydroxide ions in an aqueous solution. Arrhenius believed that in order to do so, the base must contain hydroxide in the formula. This makes the Arrhenius model limited, as it cannot explain the basic properties of aqueous solutions of ammonia (NH_3) or its organic derivatives (amines). There are also bases that do not contain a hydroxide ion but nevertheless react with water, resulting in an increase in the concentration of the hydroxide ion. An example of this is the reaction between ammonia and water to produce ammonium and hydroxide. In this reaction ammonia is the base because it accepts a proton from the water molecule. Ammonia and other bases similar to it usually have the ability to form a bond with a proton due to the unshared pair of electrons that they possess. In the more general Brønsted–Lowry acid–base theory, a base is a substance that can accept hydrogen cations (H^+)—otherwise known as protons. In the Lewis model, a base is an electron pair donor.

Properties

General properties of bases include:

- Concentrated or strong bases are caustic on organic matter and react violently with acidic substances.

- Aqueous solutions or molten bases dissociate in ions and conduct electricity.

- Reactions with indicators: bases turn red litmus paper blue, phenolphthalein pink, keep bromothymol blue in its natural colour of blue, and turn methyl orange yellow.

- The pH of a basic solution at standard conditions is greater than seven.

- Bases are bitter in taste.

Reactions between bases and Water

The following reaction represents the general reaction between a base (B) and water to produce a conjugate acid (BH⁺) and a conjugate base (OH⁻):

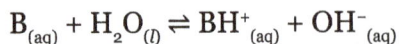

$$B_{(aq)} + H_2O_{(l)} \rightleftharpoons BH^+_{(aq)} + OH^-_{(aq)}$$

The equilibrium constant, K_b, for this reaction can be found using the following general equation:

$$K_b = [BH^+][OH^-]/[B]$$

In this equation, the base (B) and the extremely strong base (the conjugate base OH⁻) compete for the proton. As a result, bases that react with water have relatively small equilibrium constant values. The base is weaker when it has a lower equilibrium constant value.

Neutralization of Acids

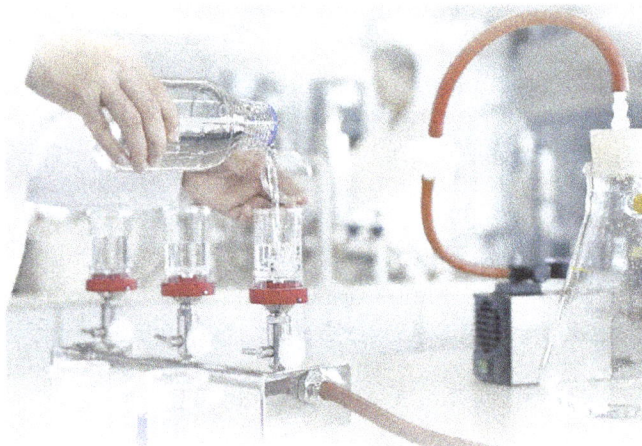

Ammonia fumes from aqueous ammonium hydroxide (in test tube) reacting with hydrochloric acid (in beaker) to produce ammonium chloride (white smoke).

Bases react with acids to neutralize each other at a fast rate both in water and in alcohol. When dissolved in water, the strong base sodium hydroxide ionizes into hydroxide and sodium ions:

$$NaOH \rightarrow Na^+ + OH^-$$

and similarly, in water the acid hydrogen chloride forms hydronium and chloride ions:

$$HCl + H_2O \rightarrow H_3O^+ + Cl^-$$

When the two solutions are mixed, the H_3O^+ and OH⁻ ions combine to form water molecules:

$$H_3O^+ + OH^- \rightarrow 2\ H_2O$$

If equal quantities of NaOH and HCl are dissolved, the base and the acid neutralize exactly, leaving only NaCl, effectively table salt, in solution.

Weak bases, such as baking soda or egg white, should be used to neutralize any acid spills. Neutralizing acid spills with strong bases, such as sodium hydroxide or potassium hydroxide, can cause a violent exothermic reaction, and the base itself can cause just as much damage as the original acid spill.

Alkalinity of Non-hydroxides

Bases are generally compounds that can neutralize an amount of acids. Both sodium carbonate and ammonia are bases, although neither of these substances contains OH^- groups. Both compounds accept H^+ when dissolved in protic solvents such as water:

$$Na_2CO_3 + H_2O \rightarrow 2\,Na^+ + HCO_3^- + OH^-$$

$$NH_3 + H_2O \rightarrow NH_4^+ + OH^-$$

From this, a pH, or acidity, can be calculated for aqueous solutions of bases. Bases also directly act as electron-pair donors themselves:

$$CO_3{}^{2-} + H^+ \rightarrow HCO_3^-$$

$$NH_3 + H^+ \rightarrow NH_4^+$$

A base is also defined as a molecule that has the ability to accept an electron pair bond by entering another atom's valence shell through its possession of one electron pair. There are a limited number of elements that have atoms with the ability to provide a molecule with basic properties. Carbon can act as a base as well as nitrogen and oxygen. Fluorine and sometimes rare gases possess this ability as well. This occurs typically in compounds such as butyl lithium, alkoxides, and metal amides such as sodium amide. Bases of carbon, nitrogen and oxygen without resonance stabilization are usually very strong, or superbases, which cannot exist in a water solution due to the acidity of water. Resonance stabilization, however, enables weaker bases such as carboxylates; for example, sodium acetate is a weak base.

Strong Bases

A strong base is a basic chemical compound that can remove a proton (H^+) from (or *deprotonate*) a molecule of even a very weak acid (such as water) in an acid-base reaction. Common examples of strong bases include hydroxides of alkali metals and alkaline earth metals, like NaOH and Ca(OH)$_2$, respectively. Due to their low solubility, some bases, such as alkaline earth hydroxides, can be used when the solubility factor is not taken into account. One advantage of this low solubility is that "many antacids were suspensions of metal hydroxides such as aluminum hydroxide and magnesium hydroxide." These compounds have low solubility and have the ability to stop an increase in the concentration of the hydroxide ion, preventing the harm of the tissues in the mouth, oesophagus, and stomach. As the reaction continues and the salts dissolve, the stomach acid reacts with the hydroxide produced by the suspensions. Strong bases hydrolyze in water almost completely, resulting in the leveling effect." In this process, the water molecule combines with a strong base, due to the water's amphoteric ability; and, a hydroxide ion is released. Very strong bases can even deprotonate very weakly acidic C–H groups in the absence of water. Here is a list of several strong bases:

- Lithium hydroxide (LiOH),

- Sodium hydroxide (NaOH),

- Potassium hydroxide (KOH),

- Rubidium hydroxide (RbOH),

- Cesium hydroxide (CsOH),

- Calcium hydroxide ($Ca(OH)_2$),

- Strontium hydroxide ($Sr(OH)_2$),

- Barium hydroxide ($Ba(OH)_2$).

The cations of these strong bases appear in the first and second groups of the periodic table (alkali and earth alkali metals).

Acids with a pK_a of more than about 13 are considered very weak, and their conjugate bases are strong bases.

Superbases

Group 1 salts of carbanions, amides, and hydrides tend to be even stronger bases due to the extreme weakness of their conjugate acids, which are stable hydrocarbons, amines, and dihydrogen. Usually these bases are created by adding pure alkali metals such as sodium into the conjugate acid. They are called *superbases*, and it is impossible to keep them in water solution because they are stronger bases than the hydroxide ion. As such, they deprotonate the conjugate acid water. For example, the ethoxide ion (conjugate base of ethanol) in the presence of water undergoes this reaction.

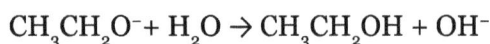

$$CH_3CH_2O^- + H_2O \rightarrow CH_3CH_2OH + OH^-$$

Examples of superbases are:

- Ortho-diethynylbenzene dianion ($C_6H_4(C_2)_2)^{2-}$ (This is the strongest Superbase ever synthesized).

- Meta-diethynylbenzene dianion ($C_6H_4(C_2)_2)^{2-}$ (This is the second strongest Superbase, this is an isomer of Ortho-diethynylbenzene dianion).

- Para-diethynylbenzene dianion ($C_6H_4(C_2)_2)^{2-}$ (This is the third strongest Superbase, this is also an isomer of Ortho-diethynylbenzene dianion).

- Lithium monoxide anion (LiO^-) (This is the fourth strongest Superbase).

- Butyl lithium (n-C_4H_9Li).

- Lithium diisopropylamide (LDA) $[(CH_3)_2CH]_2NLi$.

- Lithium diethylamide (LDEA) $(C_2H_5)_2NLi$.

- Sodium amide ($NaNH_2$).

- Sodium hydride (NaH).

- Lithium bis(trimethylsilyl)amide $[(CH_3)_3Si]_2NLi$.

Neutral Bases

When a neutral base forms a bond with a neutral acid, a condition of electric stress occurs. The acid and the base share the electron pair that formerly only belonged to the base. As a result, a high dipole moment is created, which can only be destroyed by rearranging the molecules.

Weak Bases

A weak base is one which does not fully ionize in an aqueous solution, or in which protonation is incomplete.

Solid Bases

Examples of solid bases include:

- Oxide mixtures: SiO_2, Al_2O_3; MgO, SiO_2; CaO, SiO_2.

- Mounted bases: $LiCO_3$ on silica; NR_3, NH_3, KNH_2 on alumina; $NaOH$, KOH mounted on silica on alumina.

- Inorganic chemicals: BaO, $KNaCO_3$, BeO, MgO, CaO, KCN.

- Anion exchange resins.

- Charcoal that has been treated at 900 degrees Celsius or activates with N_2O, NH_3, $ZnCl_2$-NH_4Cl-CO_2.

Depending on a solid surface's ability to successfully form a conjugate base by absorbing an electrically neutral acid, the basic strength of the surface is determined. "The number of basic sites per unit surface area of the solid" is used to express how much base is found on a solid base catalyst. Scientists have developed two methods to measure the amount of basic sites: titration with benzoic acid using indicators and gaseous acid adsorption. A solid with enough basic strength will absorb an electrically neutral acid indicator and cause the acid indicator's color to change to the color of its conjugate base. When performing the gaseous acid adsorption method, nitric oxide is used. The basic sites are then determined using the amount of carbon dioxide than is absorbed.

Bases as Catalysts

Basic substances can be used as insoluble heterogeneous catalysts for chemical reactions. Some examples are metal oxides such as magnesium oxide, calcium oxide, and barium oxide as well as potassium fluoride on alumina and some zeolites. Many transition metals make good catalysts, many of which form basic substances. Basic catalysts have been used for hydrogenations, the migration of double bonds, in the Meerwein-Ponndorf-Verley reduction, the Michael reaction, and many other reactions. Both CaO and BaO can be highly active catalysts if they are treated with high temperature heat.

Uses of Bases

- Sodium hydroxide is used in manufacture of soap, paper and the synthetic fiber rayon.

- Calcium hydroxide (slaked lime) is used in the manufacture of bleaching powder.

- Calcium hydroxide is also used to clean the sulfur dioxide, which is caused by exhaust, that is found in power plants and factories.

- Magnesium hydroxide is used as an 'antacid' to neutralize excess acid in the stomach and cure indigestion.

- Sodium carbonate is used as washing soda and for softening hard water.

- Sodium bicarbonate (or sodium hydrogen carbonate) is used as baking soda in cooking food, for making baking powders, as an antacid to cure indigestion and in soda acid fire extinguisher.

- Ammonium hydroxide is used to remove grease stains from clothes.

Acidity of Bases

The number of ionizable hydroxide (OH-) ions present in one molecule of base is called the acidity of bases. On the basis of acidity bases can be classified into three types: monoacidic, diacidic and triacidic.

Monoacidic Bases

Sodium hydroxide.

When one molecule of a base via complete ionization produces one hydroxide ion, the base is said to be a monoacidic base. Examples of monoacidic bases are:

Sodium hydroxide, potassium hydroxide, silver hydroxide, ammonium hydroxide, etc

Diacidic Bases

When one molecule of base via complete ionization produces two hydroxide ions, the base is said to be diacidic. Examples of diacidic bases are:

Barium hydroxide, magnesium hydroxide, calcium hydroxide, zinc hydroxide, iron(II) hydroxide, tin(II) hydroxide, lead(II) hydroxide, copper(II) hydroxide, etc.

Barium hydroxide.

Triacidic Bases

When one molecule of base via complete ionization produces three hydroxide ions, the base is said to be triacidic. Examples of triacidic bases are:

- Aluminium hydroxide, ferrous hydroxide, Gold Trihydroxide, etc.

Acid–base Reaction

Acid–base reaction is a type of chemical process typified by the exchange of one or more hydrogen ions, H^+, between species that may be neutral (molecules, such as water, H_2O; or acetic acid, CH_3CO_2H) or electrically charged (ions, such as ammonium, NH_4^+; hydroxide, OH^-; or carbonate, CO_3^{2-}). It also includes analogous behaviour of molecules and ions that are acidic but do not donate hydrogen ions (aluminum chloride, $AlCl_3$, and the silver ion AG^+).

Acids are chemical compounds that show, in water solution, a sharp taste, a corrosive action on metals, and the ability to turn certain blue vegetable dyes red. Bases are chemical compounds that, in solution, are soapy to the touch and turn red vegetable dyes blue. When mixed, acids and bases neutralize one another and produce salts, substances with a salty taste and none of the characteristic properties of either acids or bases.

The idea that some substances are acids whereas others are bases is almost as old as chemistry, and the terms *acid*, *base*, and *salt* occur very early in the writings of the medieval alchemists. Acids were probably the first of these to be recognized, apparently because of their sour taste. The English word *acid*, the French *acide*, the German *Säure*, and the Russian *kislota* are all derived from words meaning sour. Other properties associated at an early date with acids were their solvent, or corrosive, action; their effect on vegetable dyes; and the effervescence resulting when they were applied to chalk (production of bubbles of carbon dioxide gas). Bases (or alkalies) were characterized mainly by their ability to neutralize acids and form salts, the latter being typified rather loosely as crystalline substances soluble in water and having a saline taste.

In spite of their imprecise nature, these ideas served to correlate a considerable range of qualitative observations, and many of the commonest chemical materials that early chemists encountered could be classified as acids (hydrochloric, sulfuric, nitric, and carbonic acids), bases (soda, potash, lime, ammonia), or salts (common salt, sal ammoniac, saltpetre, alum, borax). The absence of any apparent physical basis for the phenomena concerned made it difficult to make quantitative progress in understanding acid–base behaviour, but the ability of a fixed quantity of acid to neutralize a fixed quantity of base was one of the earliest examples of chemical equivalence: the idea that a certain measure of one substance is in some chemical sense equal to a different amount of a second substance. In addition, it was found quite early that one acid could be displaced from a salt with another acid, and this made it possible to arrange acids in an approximate order of strength. It also soon became clear that many of these displacements could take place in either direction according to experimental conditions. This phenomenon suggested that acid–base reactions are reversible—that is, that the products of the reaction can interact to regenerate the starting material. It also introduced the concept of equilibrium to acid–base chemistry: this concept states that reversible chemical reactions reach a point of balance, or equilibrium, at which the starting materials and the products are each regenerated by one of the two reactions as rapidly as they are consumed by the other.

Apart from their theoretical interest, acids and bases play a large part in industrial chemistry and in everyday life. Sulfuric acid and sodium hydroxide are among the products manufactured in largest amounts by the chemical industry, and a large percentage of chemical processes involve acids or bases as reactants or as catalysts. Almost every biological chemical process is closely bound up with acid–base equilibria in the cell, or in the organism as a whole, and the acidity or alkalinity of the soil and water are of great importance for the plants or animals living in them. Both the ideas and the terminology of acid–base chemistry have permeated daily life, and the term salt is especially common.

Theoretical Definitions of Acids and Bases

Hydrogen and Hydroxide Ions

The first attempt at a theoretical interpretation of acid behaviour was made by Antoine-Laurent Lavoisier at the end of the 18th century. Lavoisier supposed that all acids must contain oxygen, and this idea was incorporated in the names used for this element in the various languages; Following the discovery that hydrochloric acid contained no oxygen, Sir Humphry Davy about 1815 first recognized that the key element in acids was hydrogen. Not all substances that contain hydrogen, however, are acids, and the first really satisfactory definition of an acid was given by Justus von Liebig of Germany in 1838. According to Liebig, an acid is a compound containing hydrogen in a form in which it can be replaced by a metal. This definition held the field for about 50 years and is still considered essentially correct, though somewhat outmoded. At the time of Liebig's proposal, bases were still regarded solely as substances that neutralized acids with the production of salts, and nothing was known about the constitutional features of bases that enabled them to do this.

The whole subject of acid–base chemistry acquired a new look and a quantitative aspect with the advent of the electrolytic dissociation theory propounded by Wilhelm Ostwald and Svante August Arrhenius (both Nobel laureates) in the 1880s. The principal feature of this theory is that certain compounds, called electrolytes, dissociate in solution to give ions. With the development of this theory it was realized that acids are merely those hydrogen compounds that give rise to hydrogen ions (H^+) in aqueous solution. It was also realized at that time that there is a correspondence between the degree of acidity of a solution (as shown by effects on vegetable dyes and other properties) and the concentration of hydrogen ions in the solution. Correspondingly, basic (or alkaline) properties could then be associated with the presence of hydroxide ions (OH^-) in aqueous solution, and the neutralization of acids by bases could be explained in terms of the reaction of these two ions to give the neutral molecule water ($H^+ + OH^- \rightarrow H_2O$). This led naturally to the simple definition that acids and bases are substances that give rise, respectively, to hydrogen and hydroxide ions in aqueous solution. This definition was generally accepted for the next 30 or 40 years. In purely qualitative terms, it does not offer many advantages over Liebig's definition of acids, but it does provide a satisfactory definition for bases.

Nevertheless, there is a great advantage in the definition of acids and bases in terms of hydrogen and hydroxide ions, and this advantage lies in its quantitative aspects. Because the concentrations of hydrogen and hydroxide ions in solution can be measured, notably by determining the electrical conductivity of the solution (its ability to carry an electrical current), a quantitative measure of the acidity or alkalinity of the solution is provided. Moreover, the equations developed to express the relationships between the various components of reversible reactions can be applied to acid and base dissociations to give definite values, called dissociation constants. These constants can be used to characterize the relative strengths (degrees of dissociation) of acids and bases and, for this reason, supersede earlier semiquantitative estimates of acid or base strength. As a result of this approach, a satisfactory quantitative description was given at an early date of a large mass of experimental observations, a description that remains essentially unaffected by later developments in definitions of acid–base reactions.

The success of these quantitative developments, however, unfortunately helped to conceal some ambiguities and logical inconsistencies in the qualitative definitions of acids and bases in terms of the production of hydrogen and hydroxide ions, respectively. For example, it was not clear whether a substance like anhydrous hydrogen chloride, which would not conduct electricity, should be regarded as an acid or whether it should be considered an acid only after it had come in contact with water. A modified definition of bases also seemed to be required that would be applicable to nonaqueous solutions, in which the anion (negatively charged ion) produced is not the hydroxide ion, as is the case in water, but varies from solvent to solvent, the methoxide ion (CH_3O^-) acting as the basic anion in methanol (CH_3OH), for example, and the amide ion (NH_2^-) playing the same role in liquid ammonia(NH_3). Even with acids the solvent is involved, since there is much evidence to show that the so-called hydrogen ion in solution does not exist as H^+ but always contains at least one molecule of solvent, as H_3O^+ in water, $CH_3OH_2^+$ in methanol, and NH_4^+ in liquid ammonia. These considerations led to the development of definitions of acids and bases that depended on the solvent. In spite of this change, however, the difficulty still remained that typical acid–base properties, such as neutralization, indicator (vegetable dye) effects, and catalysis, often took place in solvents such as benzene or chloroform in which free ions could barely be detected at all (by conductivity measurements). Even for aqueous solutions a particular ambiguity arises in the definition of bases, some of which (for example, metallic hydroxides) contain a hydroxyl group, whereas others (such as amines) do not. The latter produce hydroxide ions in solution by reacting with water molecules.

The Brønsted–Lowry Definition

In order to resolve the various difficulties in the hydrogen–hydroxide ion definitions of acids and bases, a new, more generalized definition was proposed in 1923 almost simultaneously by J.M. Brønsted and T.M. Lowry. Although the pursuit of exact verbal definitions of qualitative concepts is usually not profitable in physical science, the Brønsted–Lowry definition of acids and bases has had far-reaching consequences in the understanding of a wide range of phenomena and in the stimulation of much experimental work. The definition is as follows: an acid is a species having a tendency to lose a proton, and a base is a species having a tendency to gain a proton. The term proton means the species H^+ (the nucleus of the hydrogen atom) rather than the actual hydrogen ions that occur in various solutions; the definition is thus independent of the solvent. The use of the word species rather than substance or molecule implies that the terms acid and base are not restricted to uncharged molecules but apply also to positively or negatively charged ions. This extension is one of the important features of the Brønsted–Lowry definition. It can be summarized by the equation $A \rightleftarrows B + H^+$, in which A and B together are a conjugate acid–base pair. In such a pair A must obviously have one more positive charge (or one less negative charge) than B, but there is no other restriction on the sign or magnitude of the charges.

Several examples of conjugate acid–base pairs are given in the table.

Examples of conjugate acid-base pairs	
acid	base
acetic acid, CH_3CO_2H	acetate ion, $CH_3CO_2^-$
bisulfate ion, HSO_4^-	sulfate ion, SO_4^{2-}
ammonium ion, NH_4^+	ammonia, NH_3
ammonia, NH_3	amide ion, NH_2^-
water, H_2O	hydroxide ion, OH^-
hydronium (oxonium) ion, H_3O^+	water, H_2O

A number of points about the Brønsted–Lowry definition should be emphasized:

- As mentioned above, this definition is independent of the solvent. The ions derived from the solvent (H_3O^+ and OH^- in water and NH_4^+ and NH_2^- in liquid ammonia) are not accorded any special status but appear as examples of acids or bases in terms of the general definition. On the other hand, of course, they will be particularly important species for reactions in the solvent to which they relate.

- In addition to the familiar molecular acids, two classes of ionic acids emerge from the new definition. The first comprises anions derived from acids containing more than one acidic hydrogen—e.g., the bisulfate ion (HSO_4^-) and primary and secondary phosphate ions ($H_2PO_4^-$ and HPO_4^{2-}) derived from phosphoric acid (H_3PO_4). The second and more interesting class consists of positively charged ions (cations), such as the ammonium ion (NH_4^+), which can be derived by the addition of a proton to a molecular base, in this case ammonia (NH_3). The hydronium ion (H_3O^+), which is the hydrogen ion in aqueous solution, also belongs to this class. The charge of these ionic acids, of course, always must be balanced by ions of opposite charges, but these oppositely charged ions usually are irrelevant to the acid–base properties of the system. For example, if sodium bisulfate ($Na^+HSO_4^-$) or ammonium chloride($NH_4^+Cl^-$) is used as an acid, the sodium ion (Na^+) and the chloride ion (Cl^-) contribute nothing to the acidic properties and could equally well be replaced by other ions, such as potassium (K^+) and perchlorate (ClO_4^-), respectively.

- Molecules such as ammonia and organic amines are bases by virtue of their tendency to accept a proton. With metallic hydroxides such as sodium hydroxide, on the other hand, the basic properties are due to the hydroxide ion itself, the sodium ion serving merely to preserve electrical neutrality. Moreover, not only the hydroxide ion but also the anions of other weak acids (for example, the acetate ion) must be classed as bases because of their tendency to reform the acid by accepting a proton. Formally, the anion of any acid might be regarded as a base, but for the anion of a very strong acid (the chloride ion, for example) the tendency to accept a proton is so weak that its basic properties are insignificant and it is inappropriate to describe it as a base. Similarly, all hydrogen compounds could formally be defined as acids, but in many of them (for example, most hydrocarbons, such as methane, CH_4) the tendency to lose a proton is so small that the term *acid* would not normally be applied to them.

- Some species, including molecules as well as ions, possess both acidic and basic properties; such materials are said to be amphoteric. Both water and ammonia are amphoteric, a situation that can be represented by the schemes $H_3O^+ - H_2O - OH^-$ and $NH_4^+ - NH_3 - NH_2^-$. Another example is the secondary phosphate ion, HPO_4^{2-}, which can either lose or accept a proton, according to the following equations: $HPO_4^{2-} \rightleftarrows PO_4^{3-} + H^+$ and $HPO_4^{2-} + H^+ \rightleftarrows H_2PO_4^-$. The amphoteric properties of water are particularly important in determining its properties as a solvent for acid–base reactions.

- The equation $A \rightleftarrows B + H^+$, used in the Brønsted–Lowry definition, does not represent a reaction that can be observed in practice, since the free proton, H^+, can be observed only in gaseous systems at low pressures. In solution, the proton always is attached to some other species, commonly a solvent molecule. Thus in water the ion H_3O^+ consists of a proton bound to a water molecule. For this reason all observable acid–base reactions in solution are combined in pairs, with the result that they are of the form $A_1 + B_2 \rightleftarrows B_1 + A_2$. The fact that the process $A \rightleftarrows B + H^+$ cannot be observed does not imply any serious inadequacy of the definition. A similar situation exists with the definitions of oxidizing and reducing agents, which are defined respectively as species having a tendency to gain or lose electrons, even though one of these reactions never occurs alone and free electrons are never detected in solution (any more than free protons are).

Alternative Definitions

Although the Brønsted–Lowry concept of acids and bases as donors and acceptors of protons is still the most generally accepted one, other definitions are often encountered. Certain of these are adapted for special situations only, but the most important of these other definitions is in some respects more general than the Brønsted–Lowry definition. This definition was first proposed by the American chemist Gilbert N. Lewis in 1923.

According to Lewis, an acid is a species that can accept an electron pair from a base with the formation of a chemical bond composed of a shared electron pair (covalent bond). This classification includes as bases the same species covered by the Brønsted–Lowry definition, since a molecule or ion that can accept a proton does so because it has one or more unshared pairs of electrons, and therefore it also can combine with electron acceptors other than the proton. On the other hand, the typical Lewis acids need not (and usually do not) contain protons, being species with outer electron shells that are capable of expansion, such as boron trifluoride (BF_3), sulfur trioxide (SO_3), and silver ion (Ag^+). Lewis originally based his ideas on the experimental fact that these nonprotonic acids often exhibit the properties regarded as typical of acids, such as neutralization of bases, action on indicators, and catalysis. Such substances often are electron acceptors, but this is not always the case; carbon dioxide (CO_2) and nitrogen pentoxide (N_2O_5), for example, contain completed octets of electrons and, according to usual valence theory, cannot accept any more. In addition, hydrogen-containing substances that have always been regarded as acids (acetic acid, for example) are not obviously electron acceptors, being rather adducts of the proton (a true Lewis acid) and a base such as the acetate ion. They can only be brought logically into the Lewis scheme by appealing to the fact that the reaction between a proton acid, which may be designated as XH, and a base, denoted by B, passes through an intermediate hydrogen-bonded state, X-H...B (in which the dotted line indicates a hydrogen bond, a relatively weak secondary attractive force).

Numerous lengthy polemical exchanges have taken place regarding the relative merits of the Brønsted–Lowry and Lewis definitions. The difference is essentially one of nomenclature and has little scientific content. In the remainder of this article the term *acid* is used to denote a proton donor (following the Brønsted–Lowry terminology), whereas the term *Lewis acid* is employed exclusively to refer to electron-pair acceptors. This choice is based partly on the logical difficulties mentioned in the last paragraph and partly on the fact that the quantitative description of acid–base reactions is much simpler when it is confined to proton acids. It also represents the commonest usage of the terms.

The definition of Lewis acids and bases in terms of the gain or loss of electrons should not be confused with the definition of oxidizing and reducing agents in similar terms. In oxidation–reduction reactionsone or more electrons are transferred completely from the reducing agent to the oxidizing agent, whereas in a Lewis acid–base reaction an electron pair on the base is used to form a covalent link with the acid.

Certain other acid–base definitions have been based upon reactions occurring in specific solvent systems. For proton acids in amphoteric solvents these are equivalent to the Brønsted–Lowry definition. It is sometimes convenient to have general terms for the cation and anion derived from the solvent molecule by the addition and removal of a proton, respectively. The terms *lyonium* and *lyate ions* are occasionally used in this way. In water, the lyonium and lyate ions are H_3O^+ and OH^-; in ethanol, $C_2H_5OH_2^+$ and $C_2H_5O^-$; and in liquid ammonia, NH_4^+ and NH_2^-. For a given solvent, an acid can then be defined as a substance that increases the lyonium ion concentration (and correspondingly decreases the lyate ion concentration), whereas a base increases the lyate ion concentration (and decreases the lyonium ion concentration). This kind of definition, to be sure, really does not add anything to the concept of acids and bases as proton donors and proton acceptors.

The idea that an acid is a solute that gives rise to cations characteristic of the solvent and that a base is a solute that gives rise to anions characteristic of the solvent has sometimes been extended to solvents where no protons are involved at all—for example, liquid sulfur dioxide, SO_2. In this example, the solvent is supposed to ionize according to the equation $2SO_2 \rightleftarrows SO^{2+} + SO_3^{2-}$. Thionyl chloride, regarded as $SO^{2+} + 2Cl^-$, then can be considered an acid, and potassium sulfite, $2K^+ + SO_3^{2-}$, can be considered a base. The species SO^{2+} and SO_3^{2-} can certainly be regarded as Lewis acids and bases, but it is doubtful that they exist to any appreciable extent in liquid sulfur dioxide, a situation that makes the discussion somewhat artificial. Although this view of acids and bases has been useful in stimulating work in unusual types of solvent (for example, in carbonyl chloride, selenium oxychloride, antimony trichloride, and hydrogen cyanide), it has not met with general acceptance.

Acid–Base Reactions

Proton-transfers

The reaction expressed by the Brønsted–Lowry definition, $A \rightleftarrows B + H^+$, does not actually occur in any solution processes. This is because H^+, the bare proton, has an enormous tendency to add to almost all chemical species and cannot exist in any detectable concentrations except in a high vacuum. Apart from any specific chemical interaction, the very small size of the proton (about 10^{-15} metre) means that it exerts an extremely powerful electric field, which will polarize and therefore attract any molecule or ion it comes into contact with. It has been estimated that the dissociation of 19 grams of the hydronium ion H_3O^+ to give 1 gram of protons and 18 grams of water would require the expenditure of about 1,200,000 joules (290,000 calories) of energy, and thus it is an extremely unlikely process indeed.

Typical acid–base reactions may be thought of as the combination of two reaction schemes, $A_1 \rightleftarrows B_1 + H^+$ and $H^+ + B_2 \rightleftarrows A_2$, leading to the combined form $A_1 + B_2 \rightleftarrows B_1 + A_2$. This represents a proton-transfer reaction from A_1 to B_2, producing B_1 and A_2. A large number of reactions in solution, often referred to under a variety of names, can be represented in this way. This is illustrated by the following examples, in each of which the species are written in the order A_1, B_2, B_1, A_2.

Dissociation of Molecular Acids in Water

In this instance, water acts as a base. The equation for the dissociation of acetic acid, for example, is $CH_3CO_2H + H_2O \rightleftarrows CH_3CO_2^- + H_3O^+$.

Dissociation of Bases in Water

In this case, the water molecule acts as an acid and adds a proton to the base. An example, using ammonia as the base, is $H_2O + NH_3 \rightleftarrows OH^- + NH_4^+$. Older formulations would have written the left-hand side of the equation as ammonium hydroxide, NH_4OH, but it is not now believed that this species exists, except as a weak, hydrogen-bonded complex.

Dissociation of Acids and Bases in Nonaqueous Solvents

These situations are entirely analogous to the comparable reactions in water. For example, the dissociation of acetic acid in methanol may be written as $CH_3CO_2H + CH_3OH \rightleftarrows CH_3CO_2^- + CH_3OH$ and the dissociation of ammonia in the same solvent as $CH_3OH + NH_3 \rightleftarrows CH_3O^- + NH_4^+$.

Self-dissociation of Amphoteric Solvents

In this case, one solvent molecule acts as an acid and another as a base. Self-dissociation of water and liquid ammonia may be given as examples:

$$H_2O + H_2O \rightleftharpoons OH^- H_3O^+$$
$$NH_3 + NH_3 \rightleftharpoons NH_2^- + NH_4^+$$

Neutralization

For a strong acid and a strong base in water, the neutralization reaction is between hydrogen and hydroxide ions—i.e., $H_3O^+ + OH^- \rightleftarrows 2H_2O$. For a weak acid and a weak base, neutralization is more appropriately considered to involve direct proton transfer from the acid to the base. For example, the neutralization of acetic acid by ammonia may be written as $CH_3CO_2H + NH_3 \rightarrow CH_3CO_2^- + NH_4^+$. This equation does not involve the solvent; it therefore also represents the process of neutralization in an inert solvent, such as benzene, or in the complete absence of a solvent. (If one of the reactants is present in large excess, the reaction is more appropriately described as the dissociation of acetic acid in liquid ammonia or of ammonia in glacial acetic acid.)

Hydrolysis of Salts

Many salts give aqueous solutions with acidic or basic properties. This is termed hydrolysis, and the explanation of hydrolysis reactions in classical acid–base terms was somewhat involved. In terms of the Brønsted–Lowry concept, however, hydrolysis appears to be a natural consequence of the acidic properties of cations derived from weak bases and the basic properties of anions derived from weak acids. For example, hydrolysis of aqueous solutions of ammonium chloride and of sodium acetate is represented by the following equations:

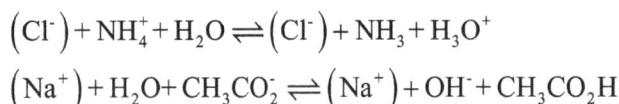

$$\left(Cl^-\right) + NH_4^+ + H_2O \rightleftharpoons \left(Cl^-\right) + NH_3 + H_3O^+$$

$$\left(Na^+\right) + H_2O + CH_3CO_2^- \rightleftharpoons \left(Na^+\right) + OH^- + CH_3CO_2H$$

The sodium and chloride ions take no part in the reaction and could equally well be omitted from the equations.

The acidity of the solution represented by the first equation is due to the presence of the hydronium ion (H_3O^+), and the basicity of the second comes from the hydroxide ion (OH^-). The reverse reactions simply represent, respectively, the neutralization of aqueous ammonia by a strong acid and of aqueous acetic acid by a strong base.

A superficially different type of hydrolysis occurs in aqueous solutions of salts of some metals, especially those giving multiply charged cations. For example, aluminum, ferric, and chromic salts all give aqueous solutions that are acidic. This behaviour also can be interpreted in terms of proton-transfer reactions if it is remembered that the ions involved are strongly hydrated in solution. In a solution of an aluminum salt, for instance, a proton is transferred from one of the water molecules in the hydration shell to a molecule of solvent water. The resulting hydronium ion (H_3O^+) accounts for the acidity of the solution:

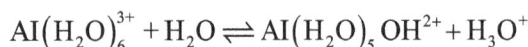

$$Al\left(H_2O\right)_6^{3+} + H_2O \rightleftharpoons Al\left(H_2O\right)_5 OH^{2+} + H_3O^+$$

Reactions of Lewis Acids

In the reaction of a Lewis acid with a base the essential process is the formation of an adduct in which the two species are joined by a covalent bond; proton transfers are not normally involved. If both the Lewis acid and base are uncharged, the resulting bond is termed semipolar or coordinate, as in the reaction of boron trifluoride with ammonia:

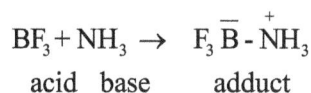

$$BF_3 + NH_3 \rightarrow F_3\overset{-}{B}\text{-}\overset{+}{N}H_3$$
$$\text{acid} \quad \text{base} \qquad \text{adduct}$$

Frequently, however, either or both species bears a charge (most commonly a positive charge on the acid or a negative charge on the base), and the location of charges within the adduct often depends upon the theoretical interpretation of the valences involved. Examples are:

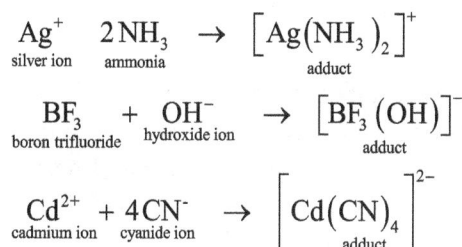

$$\underset{\text{silver ion}}{Ag^+} \quad \underset{\text{ammonia}}{2\,NH_3} \rightarrow \underset{\text{adduct}}{\left[Ag(NH_3)_2\right]^+}$$

$$\underset{\text{boron trifluoride}}{BF_3} + \underset{\text{hydroxide ion}}{OH^-} \rightarrow \underset{\text{adduct}}{\left[BF_3(OH)\right]^-}$$

$$\underset{\text{cadmium ion}}{Cd^{2+}} + \underset{\text{cyanide ion}}{4\,CN^-} \rightarrow \underset{\text{adduct}}{\left[Cd(CN)_4\right]^{2-}}$$

In another common type of process, one acid or base in an adduct is replaced by another:

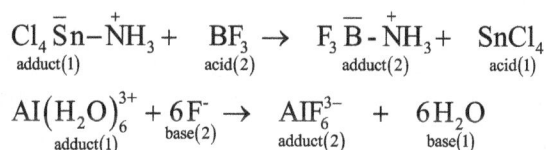

$$\underset{\text{adduct}(1)}{Cl_4\overset{-}{S}n\text{-}\overset{+}{N}H_3} + \underset{\text{acid}(2)}{BF_3} \rightarrow \underset{\text{adduct}(2)}{F_3\overset{-}{B}\text{-}\overset{+}{N}H_3} + \underset{\text{acid}(1)}{SnCl_4}$$

$$\underset{\text{adduct}(1)}{Al(H_2O)_6^{3+}} + \underset{\text{base}(2)}{6\,F^-} \rightarrow \underset{\text{adduct}(2)}{AlF_6^{3-}} + \underset{\text{base}(1)}{6\,H_2O}$$

In fact, reactions such as the simple adduct formations above often are formulated more correctly as replacements. For example, if the reaction of boron trifluoride with ammonia is carried out in ether as a solvent, it becomes a replacement reaction:

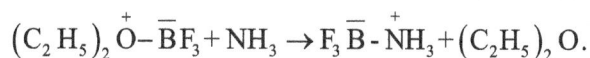

$$(C_2H_5)_2\overset{+}{O}\text{-}\overset{-}{B}F_3 + NH_3 \rightarrow F_3\overset{-}{B}\text{-}\overset{+}{N}H_3 + (C_2H_5)_2O.$$

Similarly, the reaction of silver ions with ammonia in aqueous solution is better written as a replacement reaction:

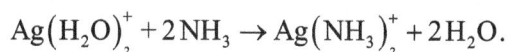

$$Ag(H_2O)_2^+ + 2\,NH_3 \rightarrow Ag(NH_3)_2^+ + 2\,H_2O.$$

Furthermore, if most covalent molecules are regarded as adducts of (often hypothetical) Lewis acids and bases, an enormous number of reactions can be formulated in the same way. To take a single example, the reaction of methyl chloride with hydroxide ion to give methanol and chloride ion (usually written as $CH_3Cl + OH^- \rightarrow CH_3OH + Cl^-$) can be reformulated as replacement of a base in a Lewis acid–base adduct, as follows: (adduct of CH_3^+ and Cl^-) + $OH^- \rightarrow$ (adduct of CH_3^+ and OH^-) + Cl^-. Opinions differ as to the usefulness of this extremely generalized extension of the Lewis acid–base-adduct concept.

The reactions of anhydrous oxides (usually solid or molten) to give salts may be regarded as examples of Lewis acid–base-adduct formation. For example, in the reaction of calcium oxide with silica to give calcium silicate, the calcium ions play no essential part in the process, which may be considered therefore to be adduct formation between silica as the acid and oxide ion as the base:

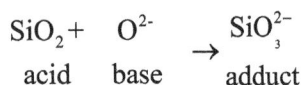

$$\underset{\text{acid}}{SiO_2} + \underset{\text{base}}{O^{2-}} \rightarrow \underset{\text{adduct}}{SiO_3^{2-}}$$

A great deal of the chemistry of molten-oxide systems can be represented in this way, or in terms of the replacement of one acid by another in an adduct.

Acid–base Catalysis

Acids (including Lewis acids) and bases act as powerful catalysts for a great variety of chemical reactions, in the laboratory, in industry, and in processes occurring in nature. Historically, catalytic action was regarded as one of the essential characteristics of acids, and the parallel occurrence of catalytic action and electrical conductivity was one of the compelling pieces of evidence in establishing the theory of electrolytic dissociation as the basis of acid–base behaviour at the end of the 19th century.

Acid–base catalysis was originally thought of in terms of a mysterious influence of the acid or base, but it is now generally believed to involve an actual acid–base reaction between the catalyst and the reacting substance, termed the substrate, with the catalyst being regenerated at a later stage of the reaction. Moreover, knowledge of reaction mechanisms is now sufficient to suggest detailed sequences of reactions for many acid- or base-catalysis reactions, most of these sequences being at least plausible and in many instances well established.

In most acid–base reactions the addition or removal of a proton does not bring about any drastic change in the structure of the molecule or in its stability or reactivity. It is a characteristic of reactions catalyzed by acids or bases, however, that the addition or removal of a proton either makes the substrate unstable, so that it decomposes or rearranges, or that it causes the substrate to become reactive toward some other species present in the system. In cases of rearrangement, the regeneration of the catalyst often involves the removal or addition of a proton at a site other than that at which the initial addition or removal took place. It is not necessary that the substrate in an acid- or base-catalyzed reaction should itself have marked acid–base properties, since even a very small extent of initial acid–base reaction may be enough to bring about the subsequent change.

Instances of acid–base catalysis are numerous indeed; a few examples are given here, as follows:

Isomerization of Olefins, Acid-catalyzed

Unsaturated compounds frequently rearrange reversibly under the influence of acids to give products in which the double bond occurs in a new location. The interconversion of 2-butene and 1-butene is shown here:

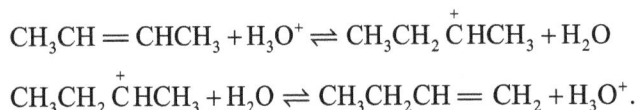

$$CH_3CH = CHCH_3 + H_3O^+ \rightleftharpoons CH_3CH_2 \overset{+}{C}HCH_3 + H_2O$$

$$CH_3CH_2 \overset{+}{C}HCH_3 + H_2O \rightleftharpoons CH_3CH_2CH = CH_2 + H_3O^+.$$

Reversible dehydration of alcohols, acid-catalyzed. Under the influence of acids, alcohols generally undergo loss of water to give olefinic products. The dehydration of ethanol to ethylene occurs as follows:

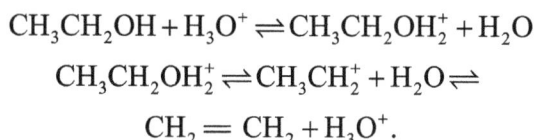

$$CH_3CH_2OH + H_3O^+ \rightleftharpoons CH_3CH_2OH_2^+ + H_2O$$
$$CH_3CH_2OH_2^+ \rightleftharpoons CH_3CH_2^+ + H_2O \rightleftharpoons$$
$$CH_2 = CH_2 + H_3O^+.$$

Keto–enol Tautomerism, Acid- and Base-catalyzed

Acids and bases both bring about the establishment of an equilibrium between ketones (or aldehydes) and their enol forms, which contain a hydroxyl group directly attached to a doubly bonded carbon atom:

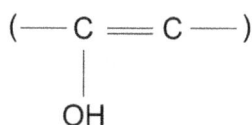

$$(-C = C-)$$
$$\quad | $$
$$\quad OH$$

The interconversion between the two forms is called keto–enol tautomerism. The reaction cannot always be observed directly, since the enol form may not reach measurable concentrations, even at equilibrium, but the highly active enol may be detected by its reaction with various reagents, notably the halogens (bromine, for example). Keto–enol tautomerization of acetone can be brought about by acid or base catalysis, as follows:

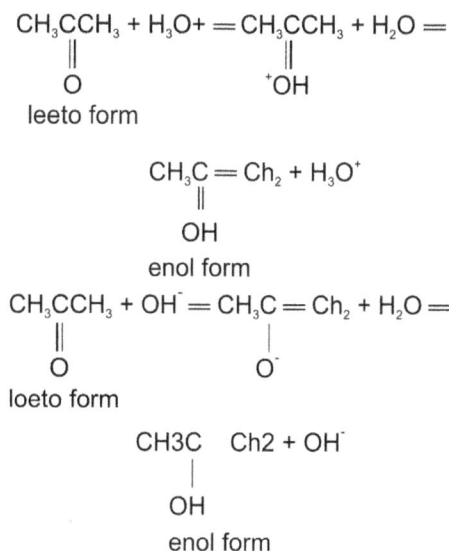

$$CH_3CCH_3 + H_3O+ = CH_3CCH_3 + H_2O =$$
$$\quad \| \qquad\qquad\qquad \|$$
$$\quad O \qquad\qquad\qquad\quad {}^+OH$$
leeto form

$$CH_3C = Ch_2 + H_3O^+$$
$$\quad \|$$
$$\quad OH$$
enol form

$$CH_3CCH_3 + OH^- = CH_3C = Ch_2 + H_2O =$$
$$\quad \| \qquad\qquad\qquad\qquad |$$
$$\quad O \qquad\qquad\qquad\qquad O^-$$
loeto form

$$CH3C \quad Ch2 + OH^-$$
$$\quad |$$
$$\quad OH$$
enol form

Aldol Condensation, Base-catalyzed

Self-condensation of aldehydes, the so-called aldol condensation, occurs readily, when catalyzed by bases, to give β-hydroxy aldehydes. The prototype of this reaction is the conversion of acetaldehyde to β-hydroxybutyraldehyde, or aldol. The first step of this reaction is the production of an enolate ion (as in formation of the keto–enol tautomeric mixture), but this anion then reacts with a second molecule of acetaldehyde to give the product as shown below:

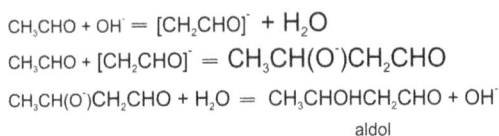

$$CH_3CHO + OH^- = [CH_2CHO]^- + H_2O$$
$$CH_3CHO + [CH_2CHO]^- = CH_3CH(O^-)CH_2CHO$$
$$CH_3CH(O^-)CH_2CHO + H_2O = CH_3CHOHCH_2CHO + OH^-$$
aldol

These examples illustrate the importance of acid–base catalysis in organic reactions. The equations have been written in terms of H_3O^+ and OH^- as the acid and base catalysts, respectively, and these are certainly the most important catalysts in aqueous solution. For many of these reactions (especially isomerization of olefins and dehydration of alcohols), there is ample evidence that other acids or bases also can act as catalysts. This behaviour is known as general acid–base catalysis. It appears particularly clearly in inert solvents such as benzene, in which catalysis by molecular acids and bases is frequently observed despite the absence of detectable quantities of ions derived from the solvent. Acidic groups, such as sulfonic acid ($-SO_3H$) and carboxylic acid groups ($-CO_2H$), attached to a solid molecular framework (as in some ion exchange resins) also act as heterogeneous catalysts for many chemical reactions.

The above examples show that proton-transfer processes can play a specific part in reaction mechanisms and, in these and similar instances, it is doubtful whether any uncatalyzed or spontaneous reaction of the same type can take place. Apparent evidence to the contrary can usually be explained by catalysis by solvent molecules or by adventitious acidic or basic impurities.

Lewis acids can exert a catalytic effect in two different ways. In the first of these they interact with hydrogen-containing compounds present in the system to assist the release of a proton to the substrate. For example, the polymerization of olefins by Lewis acids, such as boron trifluoride (BF_3), aluminum chloride ($AlCl_3$), and titanium tetrachloride ($TiCl_4$), is believed to be caused by their interaction with proton acids (for example, traces of water) and the olefin to give a carbonium ion, which then reacts further with more olefin:

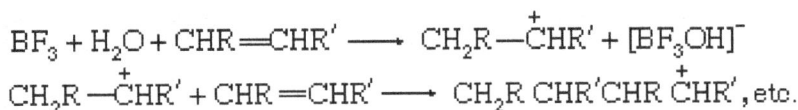

$$BF_3 + H_2O + CHR{=}CHR' \longrightarrow CH_2R{-}\overset{+}{C}HR' + [BF_3OH]^-$$
$$CH_2R{-}\overset{+}{C}HR' + CHR{=}CHR' \longrightarrow CH_2R\,CHR'CHR\,\overset{+}{C}HR', etc.$$

In the second mode of action, the Lewis acid acts directly on the substrate, and by withdrawing electrons converts it into a reactive form. A typical example is the action of catalysts like aluminum chloride on alkyl halides to produce carbonium ions: $RCl + AlCl_3 \rightarrow R^+ + [AlCl_4]^-$. The carbonium ion can then react further with other substances, for example, aromatic hydrocarbons. The same type of catalysis probably occurs with many solid oxide catalysts (for example, aluminosilicates), although it is often difficult to decide whether the catalytic action of these materials is due to centres with a deficiency of electrons or to acidic hydroxyl groups.

Acid–Base Equilibria

Certain general principles apply to any solvent with both acidic and basic properties—for example, water, alcohols, ammonia, amines, and acetic acid. Denoting the solvent molecule by SH, proton transfer can give rise to the ions SH_2^+ and S^-, sometimes called lyonium and lyate ions, respectively. In the pure solvent these are the only ions present, and they must be present in equal concentrations to preserve electrical neutrality. The equilibrium involved, therefore, is as follows: $2SH \rightleftarrows SH_2^+ + S^-$. The equilibrium constant (K_s') for this reaction (the mathematical quantity that expresses the relationships between the concentrations of the various species present at equilibrium) would normally be given by the equation $K_s' = [SH_2^+][S^-]/[SH]^2$, in which the square brackets denote the concentrations of the species within the brackets. In a given solvent, however, the concentration of the solvent, [SH], is a large and constant quantity, and it is therefore usual to eliminate this term and express the self-dissociation of the solvent by the equation $K_s = [SH_2^+][S^-]$. In this equation, K_s is termed the ion product or the autoprotolysis constant of the solvent. The concentrations are usually expressed in moles per litre, a mole being the molecular weight of the compound in grams. Since a solvent that is a good proton donor is normally a poor proton acceptor, and vice versa, the degree of ionization is generally low and K_s is usually a small quantity. It is about 10^{-14} for water at ordinary temperatures, and one of the largest K_s values known is 1.7×10^{-4} for 100 percent sulfuric acid. The above equation applies not only to the pure solvent, but also (with the same value of K_s) to any dilute solutions of acids, bases, or salts in the solvent in question. In these solutions $[SH_2^+]$ and $[S^-]$ need not be equal, since the condition of electrical neutrality involves the concentration of other ions as well, and it is obvious from the equation that a high value of $[SH_2^+]$ must imply a low value of $[S^-]$ and vice versa.

If an acid A is added to the solvent SH it will be at least partly converted into the conjugate base B according to the reaction $A + SH \rightleftarrows B + SH_2^+$, which would be characterized formally by an equilibrium constant $[B][SH_2^+]/[A][SH]$. Again, however, it is usual to omit the term for the constant concentration of the solvent, [SH], from this expression, and to define a constant K_a by the equation,

$$\frac{[B][SH_2^+]}{[A]} = K_a,$$

which is known as the dissociation constant of the acid A in the solvent SH. Any acid–base reaction $A_1 + B_2 \rightleftarrows B_1 + A_2$ will proceed from left to right almost completely if A_1 is a much stronger acid than A_2. It is a natural extension of this idea to use the equilibrium constant as a measure of the strength of A_1 relative to A_2. The dissociation constant is thus (apart from the constant factor [SH], which has been omitted) a measure of the acid strength of A relative to that of the lyonium ion SH_2^+.

In some instances reaction goes so completely from left to right that it is not possible to measure the equilibrium constant. A is said then to be a strong acid in the solvent SH; similarly, acids with readily measurable dissociation constants (in practice less than about 0.1) are known as weak acids.

Similar considerations apply to solutions of bases. The reaction involved in this case is $SH + B \rightleftarrows S^- + A$, and the equilibrium constant K_b defined by,

$$\frac{[S^-][A]}{[B]} = K_b$$

is known as the dissociation constant of the base B. Apart from the omitted constant factor [SH], K_b represents the basic strength of B relative to that of the lyate ion S^-. Bases are termed strong and weak in the same way that acids are.

The values of K_a and K_b for a conjugate acid–base pair A–B in a given solvent are not independent, since consideration of the dissociation constants of the solvent, acid, and base show that $K_a K_b = [SH_2^+][S^-] = K_s$ in which K_s is the ion product of the solvent. It is therefore unnecessary to specify both K_a and K_b, and it has become common practice to characterize an acid–base pair by K_a only, which may be termed the acidity constant of A–B in the solvent SH. If the value of K_b is required it is readily obtained from K_a and K_s. Since readily accessible values of K_a are always much less than unity, it is often convenient to introduce a quantity pK_a, sometimes called the acidity exponent, and defined by the relation $pK_a = -\log_{10} K_a$. Values of pK_a are generally of a more convenient magnitude.

The above expressions for the various equilibrium constants depend only on the concentrations of the species concerned, which are tacitly assumed to exist in solution independently of one another. This is not always the case, and in exact treatments of these equilibria two modifications are frequently necessary. In the first place, some or all of the reacting species are ions and, because of the electrical forces between them, the law expressing their concentrations at equilibrium is not always valid. Corrections may be applied by multiplying the concentrations by certain factors called activity coefficients, the values of which can be calculated theoretically or derived from other measurements. Furthermore, ions of opposite charge may attract one another so strongly that they no longer exist independently but are partly present as ion pairs, thus altering the forms of the equilibrium equations. For many purposes, however, the simple equations given here are adequate, especially with regard to reactions in aqueous solutions.

Aqueous Solutions

Since aqueous solutions are of particular importance in the laboratory and in the physiology of animals and plants, it is appropriate to consider them separately. The ion product of water, $K_w = [H_3O^+][OH^-]$, has the value 1.0×10^{-14} mole²litre⁻² at 25 °C, but it is strongly temperature-dependent, becoming 1.0×10^{-15} at 0 °C and 7×10^{-13} at 100 °C. In principle the value of K_w can be determined by measuring the electrical conductance of very pure water, in which $[H_3O^+] = [OH^-] = 10^{-7}$ at 25 °C, but in practice it is derived from other measurements—for example, measurements of the degree of hydrolysis of salts.

For an uncharged acid, in this example acetic, the dissociation constant is given by the following expression:

$$\frac{\left[CH_3CO_2^-\right]\left[H_3O^+\right]}{CH_3CO_2H} = K_a$$

For acetic acid, K_a has the value 1.76×10^{-5} at 25 °C. The dissociation constant may be expressed in terms of the degree of dissociation of the acid. This quantity, represented by the Greek letter alpha, α, is equal to the fraction of the acid that appears in dissociated form—in this case as the ions $CH_3CO_2^-$ and H_3O^+. If the initial concentration of acid is designated by c, then the concentrations of the ions are each equal to αc, or $[H_3O^+] = [CH_3CO_2^-] = \alpha c$, and the concentration of undissociated acid is equal to $c(1 - \alpha)$, or $[CH_3CO_2H] = c(1 - \alpha)$.

Substituting these expressions into the equation giving the value of the dissociation constant gives,

$$\alpha^2 c/(1-\alpha) = K_a$$

From this equation it can be inferred that the degree of dissociation (α) increases with decreasing concentration (c). For small degrees of dissociation ($\alpha << 1$), the equation becomes,

$$\alpha = K_a \left\{ \text{sup } \tfrac{1}{2} \right\} / c \left\{ \text{sup } \tfrac{1}{2} \right\}$$

whereas, at sufficiently low concentrations (c << 1), α tends to unity ($\alpha \rightarrow 1$).

Discussions exactly analogous to this apply to a number of other acid–base equilibria—for example, (1) the dissociation of ammonia in water, (2) the hydrolysis of ammonium salts, and (3) the hydrolysis of an acetate.

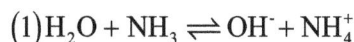

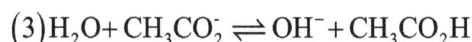

$$(1) \; H_2O + NH_3 \rightleftharpoons OH^- + NH_4^+$$

$$(2) \; NH_4^+ + H_2O \rightleftharpoons NH_3 + H_3O^+$$

$$(3) \; H_2O + CH_3CO_2^- \rightleftharpoons OH^- + CH_3CO_2H$$

For reaction (1) α is the degree of dissociation of ammonia, and the dissociation constant is K_b, the basic dissociation constant. In reaction (2), the hydrolysis of an ammonium salt (for example, ammonium chloride), α would be termed the degree of hydrolysis and K the hydrolysis constant. In terms of the general definition of acids and bases, however, K could equally be called the acidity constant for the acid–base pair NH_4^+–NH_3, and this is a more rational way of describing the process. Finally, reaction (3) represents the hydrolysis of an acetate (for example, sodium acetate); the resulting equilibrium constant is termed the hydrolysis constant and can be seen to equal K_w/K_a, where K_w is the ion product of water and K_a the acidity constant for the acid–base pair CH_3CO_2H–$CH_3CO_2^-$ (i.e., the dissociation constant of acetic acid). The investigation of equilibria such as this is, in fact, one of the methods for determining the value of K_w.

The equilibria considered so far arise when one component of an acid–base pair is dissolved in water—if necessary, along with an ion, such as Na^+ or Cl^-, having negligible acid–base properties. The direct consequence of this is that the two new species produced (for example, those on the right-hand sides of the equations [1–3] above) have equal concentrations (αc), and hence the previously given equation,

$$K = \frac{\alpha^2 c}{1 - \alpha}$$

is applicable.

A solution of a more generally useful type can be obtained by deliberately varying the proportions of acid and base present; such a solution is called a buffered solution or, somewhat more colloquially, a buffer. A buffered solution containing various concentrations of acetic acid and acetate ion, for example, can be prepared by mixing solutions of acetic acid and sodium acetate, by partially neutralizing a solution of acetic acid with sodium hydroxide, or by adding less than one equivalent of a strong acid to a solution of sodium acetate. Similarly, a buffer based on the pair NH_4^+–NH_3 can be prepared by mixing solutions of ammonia and an ammonium salt, by partially neutralizing a solution of ammonia with a strong acid, or by adding less than one equivalent of sodium hydroxide to a solution of an ammonium salt. The hydrogen ion concentration in a buffer solution is, of course, still given by the usual equation, which is conveniently written as,

$$\left[H_3O^+ \right] = K_a \frac{\left[A \right]}{\left[B \right]}$$

Since hydrogen ion concentrations are usually less than unity and cover an extremely wide range, it is often convenient to use instead the negative logarithm of the actual concentration, a figure that varies usually only in the range 1–13. The pH, and its definition is expressed by the equation pH = $-\log_{10}[H_3O^+]$. For example, in pure water $[H_3O^+] = 1 \times 10^{-7}$, with the result that the pH = 7.0. The same term can be applied to alkaline solutions; thus, in 0.1 molar sodium hydroxide $[OH^-] = 0.1$, $[H_3O^+] = K_w/[OH^-] = 1 \times 10^{-14}/0.1 = 10^{-13}$, and pH = 13.0.

Applying the pH concept to buffered solutions gives the following equation:

$$pH = pK_a - \log_{10} \frac{\left[A \right]}{\left[B \right]},$$

known as the buffer ratio, can be calculated from the way in which the solution is prepared. According to this equation, the pH of the buffered solution depends only on the pK_a of the acid and on the buffer ratio. Most particularly it does not depend on the actual concentrations of A and B. Therefore, the pH of a buffered solution is little affected by dilution of the solution. It is also insensitive to the addition of acid or alkali, provided that the amounts added are much smaller than both [A] and [B]. This so-called buffering action will be impaired if either [A] or [B] becomes too small; hence, buffer ratios must not deviate too far from unity, and the effective buffering range of a given acid–base system is roughly from pH = pK_a + 1 to pH = pK_a – 1, corresponding to buffer ratios from 0.1 to 10.

Figure shows the relation between pH and composition for a number of commonly used buffer systems. Effective buffer action is confined to the central, steep portion of each curve, where the pH is least sensitive to the composition. Figure shows that an acid bearing several acidic hydrogens, such as phosphoric acid, can be used to prepare buffer solutions in several different pH ranges. Buffer action plays an important part in controlling the pH of many biological fluids; for example, the pH of the blood is controlled at about 7.4 by the carbonic acid–bicarbonate system shown in figure. Buffers are widely used to control the pH in chemical or biological experiments. For the latter, the system $H_2PO_4^- - HPO_4^{2-}$ is particularly useful, being effective in the physiological pH range, 6–8.

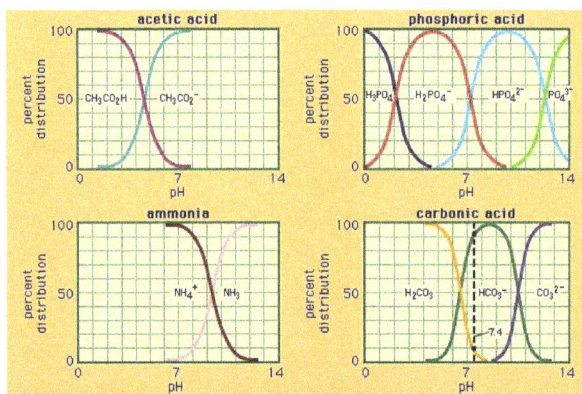

Relation between pH and composition for a number of commonly used buffer systems.

The same principles can be applied for the quantitative treatment of systems containing larger numbers of acid–base pairs; for example, in an aqueous solution of ammonium acetate, the following acid–base pairs must be considered: NH_4^+–NH_3, CH_3CO_2H–$CH_3CO_2^-$, H_3O–H_2O, and H_2O–OH^-. The situation is much more complicated in many solutions that are important in industry or in nature, but it is always possible to make a complete prediction of the state of the system in terms of the acidity constants K_a of each acid–base pair (provided, of course, that reactions other than proton transfers do not interfere).

Nonaqueous Solvents

Although acid–base properties have been investigated most thoroughly in aqueous solutions, partly because of their practical importance, water is in many respects an abnormal solvent. In particular, it has a higher dielectric constant (a measure of the ability of the medium to reduce the force between two electric charges) than most other liquids, and it is able itself to act either as an acid or as a base. The behaviour of acids and bases in several other solvents will be described briefly here.

The effect of the solvent on the dissociation of acids or bases depends largely upon the basic or acidic properties of the solvent, respectively. Since many acid–base reactions involve an increase or decrease in the number of ions, they are also influenced by the dielectric constant of the solvent, for a higher dielectric constant favours the formation of ions. Finally, the specific solvation (or close association with the solvent) of particular ions (excluding the solvation of the proton to give SH_2^+, which is already included in the basicity of the solvent) may be important. It is usually not easy to separate these three effects and, in particular, the effects of dielectric constant and solvation merge into one another. These points are illustrated with examples of several of the more important solvents. In this discussion the solvents are classified as amphoteric (both acidic and basic), acidic (in which the acidic properties are much more prominent than the basic), basic (in which the reverse is true), and aprotic (in which both acidic and basic properties are almost entirely absent). Finally, concentrated aqueous acids are mentioned as an example—a particularly important one—of mixed solvents.

Amphoteric Solvents

The most important nonaqueous solvents of this class are the lower alcohols methanol and ethanol. They resemble water in their acid–base properties but, because of their lower dielectric constants, facilitateprocesses producing ions to a much smaller extent. In particular, the ion products of these solvents are much smaller ($K_s = 10^{-17}$ for CH_3OH and 10^{-19} for C_2H_5OH, compared with 10^{-14} for water), and the dissociation constants of molecular acids and bases are uniformly lower than in water by four to five powers of 10. Nitric acid, for example, which is almost completely dissociated in water (K_a about 20), has $K_a = 2.5 \times 10^{-4}$ in methanol. On the other hand, the equilibrium constants of processes such as $NH_4^+ + ROH \rightleftarrows NH_3 + ROH_2^+$ and $CH_3CO_2^- + ROH \rightleftarrows CH_3CO_2H + RO^-$ are similar in all three solvents, since they do not involve any change in the number of ions.

Acidic Solvents

The most important strongly acidic solvent is sulfuric acid, which is able to protonate a wide variety of compounds containing oxygen or nitrogen. Thus, water, alcohols, ethers, ketones, nitro compounds, and sulfones all act as bases in sulfuric acid. This solvent must also possess some basic properties, because its ionic product is high ($[H_3SO_4^+]$ $[HSO_4^-]$ = 1.7 × 10^{-4}), but the basicity of the solvent is obscured normally by its very high acidity. For example, carboxylic acids behave as strong bases in sulfuric acid, reacting almost completely according to the equation $RCO_2H + H_2SO_4$ → $RCO_2H_2^+ + HSO_4^-$. Many substances undergo reactions in sulfuric acid that are more complicated than simple proton transfers, often yielding species important because of their chemical reactivity. Thus, some alcohols produce carbonium ions in sulfuric acid; with triphenylcarbinol, for example, the reaction is $(C_6H_5)_3COH + 2H_2SO_4$ → $(C_6H_5)_3C^+ + H_3O^+ + 2HSO_4^-$. Nitric acid gives the nitronium ion, NO_2^+, according to the equation $HNO_3 + 2H_2SO_4$ → $NO_2^+ + H_3O^+ + 2HSO_4^-$. This ion frequently is the active agent in the nitration of organic compounds. Hydrogen fluoride has solvent properties resembling those of sulfuric acid but is less acidic and has negligible basic properties. Acetic acid is another acidic solvent that has been extensively studied. Because of its low dielectric constant, ions exist in it largely in the form of ion pairs, and more complex associates are frequently formed. For this reason a quantitative interpretation of acid–base equilibria in acetic acid is often difficult, but some general conclusions can be drawn. In particular, it can be seen that all substances more basic in water solution than aniline react completely with acetic acid according to the equation $B + CH_3CO_2H$ → $BH^+ + CH_3CO_2^-$. All such bases therefore give solutions with indistinguishable acid–base properties; this is often referred to as a levelling effect of the solvent. The converse is true for acids; for example, the strong mineral acids, nitric, hydrochloric, sulfuric, hydrobromic, and perchloric (HNO_3, HCl, H_2SO_4, HBr, and $HClO_4$) are "levelled" in aqueous solution by complete conversion to the hydronium ion, but in acetic acid they are differentiated as weak acids with strengths in the approximate ratio 1:9:30:160:400.

Basic Solvents

The only basic solvent that has been investigated in any detail is liquid ammonia, which has the very low ion product $[NH_4^+]$ $[NH_2^-]$ = 10^{-33}. As might be expected, this solvent has a marked levelling effect upon acids; thus, for example, acetic, benzoic, nitric, and hydrochloric acids all give solutions with identical acidic properties, owing to the ion NH_4^+, although, of course, in water they behave very differently.

Aprotic Solvents

Strictly aprotic solvents include the hydrocarbons and their halogen derivatives, which undergo no reaction with added acids or bases. Acid–base equilibrium in these solvents can be investigated only when a second acid–base system is added; the usual reaction $A_1 + B_2 \rightleftarrows B_1 + A_2$ then takes place. Most such investigations have employed an indicator as one of the reacting systems, but the results are often difficult to interpret because of association of both ions and molecules in these media of low dielectric constant.

The term *aprotic* has been extended recently to include solvents that are unable to lose a proton, although they may have weakly basic properties. Some of these aprotic solvents have high dielectric constants (for example, N, N-dimethylformamide, dimethyl sulfoxide, and nitrobenzene) and are good solvents for a variety of substances. They have a powerful differentiating effect on the properties of acids and bases. In particular, basic anions are poorly solvated in these solvents and thus behave as very strong bases; for example, it has been estimated that sodium methoxide dissolved in dimethyl sulfoxide gives a solution 10^9 times as basic as in methanol.

Concentrated Aqueous Acids

Dilute solutions of strong acids—for example, hydrochloric, sulfuric, and perchloric (HCl, H_2SO_4, $HClO_4$)—in water behave essentially as solutions of the ion H_3O^+, and their acidity increases in proportion to their concentration. At concentrations greater than about one molar (that is, one mole of acid per litre of solution), however, the acidity, as measured by action on indicators or by catalytic ability, increases much more rapidly than the concentration. For example, a 10 molar solution of any strong acid is about 1,000 times as acidic as a 1 molar solution. This behaviour is undoubtedly largely due to the depletion of water with increasing concentration of acid; the hydronium ion, H_3O^+, is known to have a strong tendency to further hydration, probably mainly to the ion $H_3O^+(H_2O)_3$ (that is, $H_9O_4^+$), and a decrease in water content increases the proton-donating power of the solution. The acidity of these concentrated solutions is commonly measured by the acidity function, H_0, a quantity measured by the effect of the solvent on a basic indicator I. It is defined by $H_0 + pK_{IH}^+ - \log_{10}$ [IH$^+$]/[I] and becomes equal to the pH in dilute solution. The acidity function H_0 frequently is found to be independent of the nature of the indicator and to give an approximate measure of the catalytic power of the acid solution. Mixtures of sulfuric acid and water ranging from 10 to 100 percent sulfuric acid have H_0 values between −0.3 and −11.1, which corresponds to an acidity range of nearly 11 powers of 10.

Lewis Acids

Much less information is available about Lewis acid–base equilibria than about ordinary acid–base equilibria, but it is clear that the situation is less simple for the former than for the latter. When a given Lewis acid reacts with a series of similarly constituted bases the equilibrium constants often vary in parallel with the conventional basic strengths. This is the case when a zinc halide, ZnX_2, for example, reacts with a series of amines. In general, however, it is not possible to arrange Lewis acids and bases in a unique order that will predict the extent to which a given pair will react. Thus, although the hydroxide ion (OH$^-$) is always a much stronger base than ammonia (NH_3) in reactions with proton acids, in reactions with the Lewis acid Ag$^+$, the complex $Ag(NH_3)_2^+$ is fairly stable, whereas AgOH is completely dissociated. Similarly, for some metal cations complex formation increases in the order fluoride < chloride < bromide < iodide, whereas for other metal cations the order is the reverse of this.

This kind of behaviour has led to a classification of Lewis acids and bases into "hard" and "soft" categories; as a rule, hard acids react preferentially with hard bases and, similarly, soft acids react with soft bases. The terms *hard* and *soft* are chosen to suggest that the atomic structures associated with hard acids and bases are rigid and impenetrable, whereas those associated with soft acids and bases are more readily deformable. Hard acids include the proton; sodium, calcium, and aluminum ions; and carbonium ions. The soft acids include cuprous, silver, mercurous, and the halogen cations. Typical soft bases are iodide, thiocyanate, sulfide, and triphenylphosphine; whereas hard bases include hydroxide, fluoride, and many oxyanions. The dividing line between the hard and soft categories is not a sharp one, and its theoretical interpretation is obscure. Nevertheless, a surprising amount of factual information can be coordinated on the basis of preferential reactions of hard acids with hard bases and soft acids with soft bases.

The Effect of Molecular Structure

Regularities and trends in the properties of the elements are best understood in terms of the periodic table, an orderly pattern seen when the elements are arranged in order of increasing atomic number. Comparing the hydrides of the various elements in the table reveals appreciable acidity only among those of the elements on the right-hand side of the table, especially the halogen elements—fluorine, chlorine, bromine, and iodine. This generalization is borne out when the elements across the first period of the table are examined in order; as the right-hand side of the table is approached, the elements encountered are carbon, nitrogen, oxygen, and fluorine. The hydrides of these elements show increasing acidity. Methane, CH_4, a hydride of carbon, has no detectable acidic properties, and the pK_a decreases sharply in the series ammonia (NH_3), 35; water (H_2O), 16; and hydrogen fluoride (HF), 4. In any given group of the periodic table, the acidity of the hydrides increases as the group is descended. For example, the two groups at the right-hand side of the table include, respectively and in descending order, the elements oxygen, sulfur, selenium, and tellurium; and fluorine, chlorine, bromine, and iodine. The pK_a's of the hydrides of the first group are as follows: water (H_2O), 16; hydrogen sulfide (H_2S), 7; hydrogen selenide (H_2Se), 4; and hydrogen telluride (H_2Te), 3. Similarly, hydrogen fluoride (HF) is a weak acid, whereas hydrogen chloride (HCl), hydrogen bromide (HBr), and hydrogen iodide (HI) are all completely dissociated (are strong acids) in aqueous solution. These trends are due to variations in bond strength, electronegativity (attractive power of the atomic nucleus for electrons), and ionic solvation energy, of which the first is the most important. When a hydride is able to lose two or more protons, the loss of the second is always more difficult because of the increased negative charge on the base—e.g., $H_2S – HS^-$ (pK 7), $HS^- – S^{2-}$ (pK 15); similarly, $NH_4^+ – NH_3$ (pK 9.5), $NH_3 – NH_2^-$ (pK 35).

A simple rule applies to the strengths of the oxyacids, which can be given the general formula $XOn(OH)m$, in which X is any nonmetal. In these compounds, the pK decreases with increasing n but does not depend significantly upon m. When n = 0 (e.g., ClOH, Si(OH)$_4$), pK_a is between 8 and 11; when n = 1 (e.g., HNO_2, H_2SO_3) gives pK_a 2–4; whereas with n = 2 or 3 (e.g., H_2SO_4, $HClO_4$) the acids are completely dissociated in water (pK_a < 0). These regularities are probably attributable to the sharing of the negative charge of the anion between n + 1 equivalent oxygen atoms; the more extensive the charge spread, the lower is the energy of the anion and hence the stronger the acid.

The most important groups of organic acids are the alcohols (including the phenols) and the carboxylic acids. The simple alcohols are very weak acids (pK 16–19); the phenols are considerably stronger ($pK \sim$ 10); and the carboxylic acids stronger still ($pK \sim$ 5). The strength of the carboxylic acids is due to the sharing of the negative charge between two equivalent oxygen atoms in the ion RCO_2^-. The most important organic bases are the amines, RNH_2, R_2NH, or R_3N. Most of these are stronger bases than ammonia; i.e., their cations are weaker acids than the ammonium ion.

The effect of substituents on the acid–base properties of organic molecules has been very extensively studied and is one of the main methods of investigating the nature of the electron displacements produced by substitution in these molecules. The simplest classification is into electron-attracting substituents (halogens, carbonyl, nitro, and positively charged groups) and electron-repelling groups (alkyl groups, negatively charged groups). The electron-attracting groups make acids stronger and bases weaker, whereas electron-repelling groups have the opposite effects. There are, however, often more specific electronic effects, especially in aromatic and unsaturated compounds, for which special explanations are needed.

Dissociation Constants in Aqueous Solution

The classical method for determining the dissociation constant of an acid or a base is to measure the electrical conductivity of solutions of varying concentrations. From these the degree of dissociation can be determined and K_a calculated from the equation,

$$K_a = \frac{a^2 c}{1 - a}$$

This method is unsuitable for acids with pK less than 2 because a is then close to unity and the value $1 - a$ is therefore subject to error. It also is unsuitable for acids of pK > 7 because impurities in the solvent may affect the conductivity or displace the dissociation equilibrium.

It is often preferable to use a more specific method for determining the concentration of one of the species in the scheme $A + H_2O \rightleftarrows B + H_3O^+$. For example, a hydrogen electrode (or more commonly a glass electrode, which responds in the same way) together with a reference electrode, commonly the calomel electrode, serves to measure the actual hydrogen ion concentration, or the pH, of the solution. If E is the electromotive force (in volts) observed by the electrode, the equation giving the pH is as follows:

$$E = E_0 + 0.059 \log_{10} \left[H_3O^+ \right] = E_0 - 0.059 \, p\mathrm{H}.$$

In this equation, the value of E_0 depends on the nature of the reference electrode and is usually obtained by calibration with a solution of known pH. Measurements can be made on aqueous solutions of the acid, in which case $[B] = [H_3O^+]$, but it is better to use a series of buffer solutions with known ratios $[A]/[B]$, since these are less sensitive to the presence of impurities. Such a series is obtained by successive additions of alkali to a solution of the acid (or of a strong acid to a solution of the base) and the procedure is then often termed a pH titration.

If A and B have different optical properties—for example, if they differ in colour or in the absorption of ultraviolet light—this property can be used to measure the ratio $[A]/[B]$, commonly by using an instrument called a spectrophotometer. Since $[H_3O^+]$ must also be known, the commonest procedure is to measure $[A]/[B]$ in a solution made by adding a small quantity of A or B to a standard buffer solution. If A and B do not have convenient optical properties—as is commonly the case—an indicator, that is, an acid–base system that does show a difference in colour in changing from A to B, is used. If a small quantity of indicator AI–BI, with acidity constant KI, is added to a buffer solution A–B, it is easily shown that the following relation holds:

$$K_A = K_I \frac{[B]}{[A]} \cdot \frac{[A_I]}{[B_I]},$$

in which $[A_I]/[B_I]$ is measured spectrophotometrically, and all the other quantities on the right-hand side of the equation are known.

If accurate values of K are required, it is necessary in all the above methods to take into account the effect of interionic forces upon the equation and the quantities measured. This factor can induce a considerable degree of complexity into the problem.

Selected Values of Acidity Constants

The table contains acidity constants for selected substances. These are listed as acids or bases according to the nature of the uncharged species, but in each case the value given is pK_a for the acid form (pK_a and pK_b for a conjugate acid–base pair being related by the equation,

$$pK_a + pK_b = pK_w = 14.00$$

pK$_a$'s of representative acids and bases	
Inorganic acids	
Boric acid	9.1 (20 °C)
Hypochlorous acid	7.53 (18 °C)
Hydrogen sulfide	7.0, 11.9 (18 °C)
Carbonic acid	6.4, 10.3
Phosphoric acid	2.1, 7.2, 12.8
Sulfurous acid	1.8, 6.9 (18 °C)
Nitric acid	−1.6
Sulfuric acid	(−3), 1.9
Hydrogen chloride	(−7)
Perchloric acid	(−8)
Inorganic bases	
Ammonia	9.25
Hydrazine	−0.9, 8.23 (20 °C)
Hydroxylamine	6.03 (20 °C)
Alcohols and phenols	
Methanol	15.5
Trifluoroethanol	12.37
Phenol	9.89
o-nitrophenol	7.17
m-nitrophenol	8.28
p-nitrophenol	7.15
o-chlorophenol	8.49
m-chlorophenol	8.85
p-chlorophenol	9.18
Picric acid	0.38
Carboxylic acids	
Formic	3.75 (20 °C)
Acetic	4.75
Chloroacetic	2.85
Dichloroacetic	1.48
Trichloroacetic	0.70
Oxalic	1.23, 4.19
Malonic	2.83, 5.69
Benzoic	4.19

Nitrogen bases	
Methylamine	10.66
Dimethylamine	10.73
trimethylamine	9.81
piperidine	11.12
aniline	4.63
pyridine	5.25
quinoline	4.90 (20 °C)
pyrrole	−0.27

for aqueous solutions at 25 °C). For instances in which several values of pK_a are given, these relate to successive dissociations; e.g., for phosphoric acid, they correspond to dissociations of H_3PO_4, $H_2PO_4^-$, and HPO_4^{2-}. All values given refer to aqueous solutions at or near 25 °C; parentheses indicate values that have been estimated indirectly or are uncertain for other reasons.

pH

In chemistry, pH is a scale used to specify how acidic or basic a water-based solution is. Acidic solutions have a lower pH, while basic solutions have a higher pH. At room temperature (25 °C), pure water is neither acidic nor basic and has a pH of 7.

The pH scale is logarithmic and approximates the negative of the base 10 logarithm of the molar concentration (measured in units of moles per liter) of hydrogen ions in a solution. More precisely it is the negative of the base 10 logarithm of the activity of the hydrogen ion. At 25 °C, solutions with a pH less than 7 are acidic and solutions with a pH greater than 7 are basic. The neutral value of the pH depends on the temperature, being lower than 7 if the temperature increases. The pH value can be less than 0 for very strong acids, or greater than 14 for very strong bases.

The pH scale is traceable to a set of standard solutions whose pH is established by international agreement. Primary pH standard values are determined using a concentration cell with transference, by measuring the potential difference between a hydrogen electrode and a standard electrode such as the silver chloride electrode. The pH of aqueous solutions can be measured with a glass electrode and a pH meter, or a color-changing indicator. Measurements of pH are important in chemistry, agronomy, medicine, water treatment, and many other applications.

Measurement of pH

pH is defined as the decimal logarithm of the reciprocal of the hydrogen ion activity, $a_{H}+$, in a solution.

$$pH = -\log_{10}(a_{H^+}) = \log_{10}\left(\frac{1}{a_{H^+}}\right)$$

For example, for a solution with a hydrogen ion activity of 5×10^{-6} (at that level, this is essentially the number of moles of hydrogen ions per liter of solution) there is $1/(5\times10^{-6}) = 2\times10^5$, thus such a solution has a pH of $\log_{10}(2\times10^5) = 5.3$. For a commonplace example based on the facts that the masses of a mole of water, a mole of hydrogen ions, and a mole of hydroxide ions are respectively 18 g, 1 g, and 17 g, a quantity of 10^7 moles of pure (pH 7) water, or 180 tonnes (18×10^7 g), contains close to 1 g of dissociated hydrogen ions (or rather 19 g of H_3O^+ hydronium ions) and 17 g of hydroxide ions.

Note that pH depends on temperature. For instance at 0 °C the pH of pure water is 7.47. At 25 °C it's 7.00, and at 100 °C it's 6.14.

This definition was adopted because ion-selective electrodes, which are used to measure pH, respond to activity. Ideally, electrode potential, E, follows the Nernst equation, which, for the hydrogen ion can be written as,

$$E = E^0 + \frac{RT}{F}\ln(a_{H+}) = E^0 - \frac{2.303RT}{F}pH$$

where E is a measured potential, E^0 is the standard electrode potential, R is the gas constant, T is the temperature in kelvins, F is the Faraday constant. For H^+ number of electrons transferred is one. It follows that electrode potential is proportional to pH when pH is defined in terms of activity. Precise measurement of pH is presented in International Standard ISO 31-8 as follows: A galvanic cell is set up to measure the electromotive force (e.m.f.) between a reference electrode and an electrode sensitive to the hydrogen ion activity when they are both immersed in the same aqueous solution. The reference electrode may be a silver chloride electrode or a calomel electrode. The hydrogen-ion selective electrode is a standard hydrogen electrode.

Reference electrode | concentrated solution of KCl || test solution | H_2 | Pt

Firstly, the cell is filled with a solution of known hydrogen ion activity and the emf, E_S, is measured. Then the emf, E_X, of the same cell containing the solution of unknown pH is measured.

$$pH(X) = pH(S) + \frac{E_S - E_X}{z}$$

The difference between the two measured emf values is proportional to pH. This method of calibration avoids the need to know the standard electrode potential. The proportionality constant, $1/z$ is ideally equal to $\dfrac{1}{2.303RT/F}$ the "Nernstian slope".

To apply this process in practice, a glass electrode is used rather than the cumbersome hydrogen electrode. A combined glass electrode has an in-built reference electrode. It is calibrated against buffer solutions of known hydrogen ion activity. IUPAC has proposed the use of a set of buffer solutions of known H^+ activity. Two or more buffer solutions are used in order to accommodate the fact that the "slope" may differ slightly from ideal. To implement this approach to calibration, the electrode is first immersed in a standard solution and the reading on a pH meter is adjusted to be equal to the standard buffer's value. The reading from a second standard buffer solution is then adjusted, using the "slope" control, to be equal to the pH for that solution. Further details, are given in the IUPAC recommendations. When more than two buffer solutions are used the electrode is calibrated by fitting observed pH values to a straight line with respect to standard buffer values.

Commercial standard buffer solutions usually come with information on the value at 25 °C and a correction factor to be applied for other temperatures.

The pH scale is logarithmic and therefore pH is a dimensionless quantity.

p[H]

This was the original definition of Sørensen, which was superseded in favor of pH in 1909. [H] is the concentration of hydrogen ions, denoted [H⁺] in modern chemistry, which appears to have units of concentration. More correctly, the thermodynamic activity of H⁺ in dilute solution should be replaced by $[H^+]/c_o$, where the standard state concentration c_o = 1 mol/L. This ratio is a pure number whose logarithm can be defined.

However, it is possible to measure the concentration of hydrogen ions directly, if the electrode is calibrated in terms of hydrogen ion concentrations. One way to do this, which has been used extensively, is to titrate a solution of known concentration of a strong acid with a solution of known concentration of strong alkaline in the presence of a relatively high concentration of background electrolyte. Since the concentrations of acid and alkaline are known, it is easy to calculate the concentration of hydrogen ions so that the measured potential can be correlated with concentrations. The calibration is usually carried out using a Gran plot. The calibration yields a value for the standard electrode potential, E^o, and a slope factor, f, so that the Nernst equation in the form,

$$E = E^0 + f\frac{2.303RT}{F}\log[H^+]$$

can be used to derive hydrogen ion concentrations from experimental measurements of E. The slope factor, f, is usually slightly less than one. A slope factor of less than 0.95 indicates that the electrode is not functioning correctly. The presence of background electrolyte ensures that the hydrogen ion activity coefficient is effectively constant during the titration. As it is constant, its value can be set to one by defining the standard state as being the solution containing the background electrolyte. Thus, the effect of using this procedure is to make activity equal to the numerical value of concentration.

The glass electrode (and other ion selective electrodes) should be calibrated in a medium similar to the one being investigated. For instance, if one wishes to measure the pH of a seawater sample, the electrode should be calibrated in a solution resembling seawater in its chemical composition, as detailed below.

The difference between p[H] and pH is quite small. It has been stated that pH = p[H] + 0.04. It is common practice to use the term "pH" for both types of measurement.

pH Indicators

Indicators may be used to measure pH, by making use of the fact that their color changes with pH. Visual comparison of the color of a test solution with a standard color chart provides a means to measure pH accurate to the nearest whole number. More precise measurements are possible if the color is measured spectrophotometrically, using a colorimeter or spectrophotometer. Universal indicator consists of a mixture of indicators such that there is a continuous color change from about pH 2 to pH 10. Universal indicator paper is made from absorbent paper that has been impregnated with universal indicator. Another method of measuring pH is using an electronic pH meter.

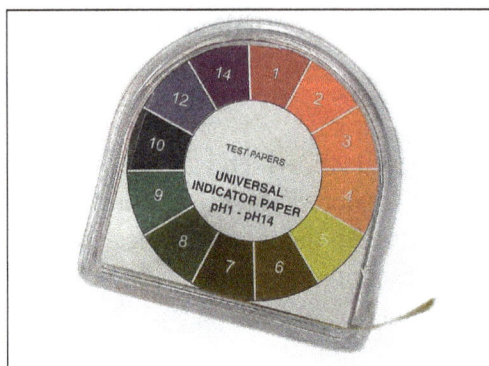

Chart showing the variation of color of universal indicator paper with pH.

pOH

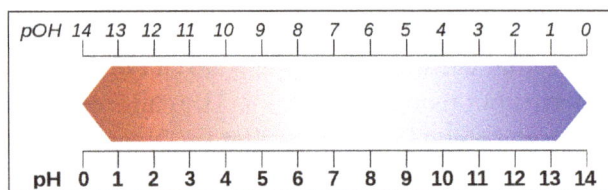

Relation between p[OH] and p[H] (red = acidic region, blue = basic region).

pOH is sometimes used as a measure of the concentration of hydroxide ions. OH⁻. pOH values are derived from pH measurements. The concentration of hydroxide ions in water is related to the concentration of hydrogen ions by,

$$[OH^-] = \frac{K_W}{[H^+]}$$

where K_W is the self-ionisation constant of water. Taking logarithms,

$$pOH = pK_W - pH$$

So, at room temperature, pOH ≈ 14 − pH. However this relationship is not strictly valid in other circumstances, such as in measurements of soil alkalinity.

Extremes of pH

Measurement of pH below about 2.5 (ca. 0.003 mol dm⁻³ acid) and above about 10.5 (ca. 0.0003 mol dm⁻³ alkaline) requires special procedures because, when using the glass electrode, the Nernst law breaks down under those conditions. Various factors contribute to this. It cannot be assumed that liquid junction potentials are independent of pH. Also, extreme pH implies that the solution is concentrated, so electrode potentials are affected by ionic strength variation. At high pH the glass electrode may be affected by "alkaline error", because the electrode becomes sensitive to the concentration of cations such as Na^+ and K^+ in the solution. Specially constructed electrodes are available which partly overcome these problems.

Runoff from mines or mine tailings can produce some very low pH values.

Non-aqueous Solutions

Hydrogen ion concentrations (activities) can be measured in non-aqueous solvents. pH values based on these measurements belong to a different scale from aqueous pH values, because activities relate to different standard states. Hydrogen ion activity, a_{H^+}, can be defined as:

$$a_{H^+} = \exp\left(\frac{\mu_{H^+} - \mu_{H^+}^{\ominus}}{RT}\right)$$

where μ_{H^+} is the chemical potential of the hydrogen ion, $\mu_{H^+}^{\ominus}$ is its chemical potential in the chosen standard state, R is the gas constant and T is the thermodynamic temperature. Therefore, pH values on the different scales cannot be compared directly due to different solvated proton ions such as lyonium ions, requiring an intersolvent scale which involves the transfer activity coefficient of hydronium/lyonium ion.

pH is an example of an acidity function. Other acidity functions can be defined. For example, the Hammett acidity function, H_o, has been developed in connection with superacids.

Unified Absolute pH Scale

The concept of "unified pH scale" has been developed on the basis of the absolute chemical potential of the proton. This model uses the Lewis acid–base definition. This scale applies to liquids, gases and even solids. In 2010, a new "unified absolute pH scale" has been proposed that would allow various pH ranges across different solutions to use a common proton reference standard.

Applications

Pure water is neutral. When an acid is dissolved in water, the pH will be less than 7 (25 °C). When a base, or alkali, is dissolved in water, the pH will be greater than 7. A solution of a strong acid, such as hydrochloric acid, at concentration 1 mol dm^{-3} has a pH of 0. A solution of a strong alkali, such as sodium hydroxide, at concentration 1 mol dm^{-3}, has a pH of 14. Thus, measured pH values will lie mostly in the range 0 to 14, though negative pH values and values above 14 are entirely possible. Since pH is a logarithmic scale, a difference of one pH unit is equivalent to a tenfold difference in hydrogen ion concentration.

The pH of neutrality is not exactly 7 (25 °C), although this is a good approximation in most cases. Neutrality is defined as the condition where $[H^+] = [OH^-]$ (or the activities are equal). Since self-ionization of water holds the product of these concentration $[H^+] \times [OH^-] = K_w$, it can be seen that at neutrality $[H^+] = [OH^-] = \sqrt{K_w}$, or pH = $pK_w/2$. pK_w is approximately 14 but depends on ionic strength and temperature, and so the pH of neutrality does also. Pure water and a solution of NaCl in pure water are both neutral, since dissociation of water produces equal numbers of both ions. However the pH of the neutral NaCl solution will be slightly different from that of neutral pure water because the hydrogen and hydroxide ions' activity is dependent on ionic strength, so K_w varies with ionic strength.

If pure water is exposed to air it becomes mildly acidic. This is because water absorbs carbon dioxide from the air, which is then slowly converted into bicarbonate and hydrogen ions (essentially creating carbonic acid).

$$CO_2 + H_2O \rightleftharpoons HCO_3^- + H^+$$

pH in Soil

Classification of Soil pH Ranges

The United States Department of Agriculture Natural Resources Conservation Service, formerly Soil Conservation Service classifies soil pH ranges as follows:

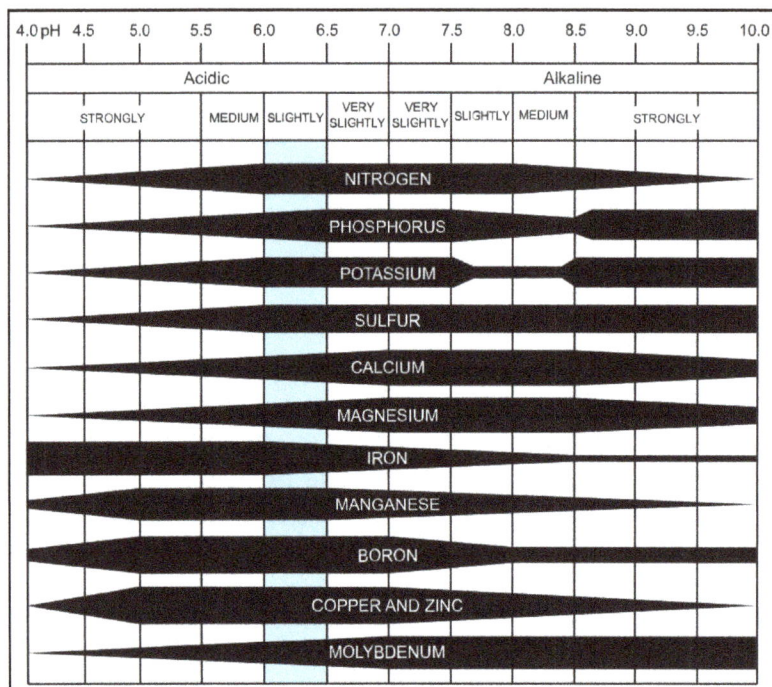

Nutritional elements availability within soil varies with pH. Light blue color represents the ideal range for most plants.

Denomination	pH range
Ultra acidic	< 3.5
Extremely acidic	3.5–4.4
Very strongly acidic	4.5–5.0
Strongly acidic	5.1–5.5
Moderately acidic	5.6–6.0
Slightly acidic	6.1–6.5
Neutral	6.6–7.3
Slightly alkaline	7.4–7.8
Moderately alkaline	7.9–8.4
Strongly alkaline	8.5–9.0
Very strongly alkaline	> 9.0

pH in Nature

Lemon juice tastes sour because it contains 5% to 6% citric acid and has a pH of 2.2. (high acidity).

pH-dependent plant pigments that can be used as pH indicators occur in many plants, including hibiscus, red cabbage (anthocyanin) and red wine. The juice of citrus fruits is acidic mainly because it contains citric acid. Other carboxylic acids occur in many living systems. For example, lactic acid is produced by muscle activity. The state of protonation of phosphate derivatives, such as ATP, is pH-dependent. The functioning of the oxygen-transport enzyme hemoglobin is affected by pH in a process known as the Root effect.

Seawater

The pH of seawater is typically limited to a range between 7.5 and 8.4. It plays an important role in the ocean's carbon cycle, and there is evidence of ongoing ocean acidification caused by carbon dioxide emissions. However, pH measurement is complicated by the chemical properties of seawater, and several distinct pH scales exist in chemical oceanography.

As part of its operational definition of the pH scale, the IUPAC defines a series of buffer solutions across a range of pH values (often denoted with NBS or NIST designation). These solutions have a relatively low ionic strength (≈ 0.1) compared to that of seawater (≈ 0.7), and, as a consequence, are not recommended for use in characterizing the pH of seawater, since the ionic strength differences cause changes in electrode potential. To resolve this problem, an alternative series of buffers based on artificial seawater was developed. This new series resolves the problem of ionic strength differences between samples and the buffers, and the new pH scale is referred to as the 'total scale', often denoted as pH_T. The total scale was defined using a medium containing sulfate ions. These ions experience protonation, $H^+ + SO_4^{2-} \rightleftharpoons HSO_4^-$, such that the total scale includes the effect of both protons (free hydrogen ions) and hydrogen sulfate ions:

$$[H^+]_T = [H^+]_F + [HSO_4^-]$$

An alternative scale, the 'free scale', often denoted 'pH_F', omits this consideration and focuses solely on $[H^+]_F$, in principle making it a simpler representation of hydrogen ion concentration. Only $[H^+]_T$ can be determined, therefore $[H^+]_F$ must be estimated using the $[SO_4^{2-}]$ and the stability constant of HSO_4^-, K_S^*:

$$[H^+]_F = [H^+]_T - [HSO_4^-] = [H^+]_T \,(1 + [SO_4^{2-}] / K_S^*)^{-1}$$

However, it is difficult to estimate K_S^* in seawater, limiting the utility of the otherwise more straightforward free scale.

Another scale, known as the 'seawater scale', often denoted 'pH_{SWS}', takes account of a further protonation relationship between hydrogen ions and fluoride ions, $H^+ + F^- \rightleftharpoons HF$. Resulting in the following expression for $[H^+]_{SWS}$:

$$[H^+]_{SWS} = [H^+]_F + [HSO_4^-] + [HF]$$

However, the advantage of considering this additional complexity is dependent upon the abundance of fluoride in the medium. In seawater, for instance, sulfate ions occur at much greater concentrations (>400 times) than those of fluoride. As a consequence, for most practical purposes, the difference between the total and seawater scales is very small.

The following three equations summarise the three scales of pH:

$$pH_F = -\log [H^+]_F$$

$$pH_T = -\log ([H^+]_F + [HSO_4^-]) = -\log [H^+]_T$$

$$pH_{SWS} = -\log ([H^+]_F + [HSO_4^-] + [HF]) = -\log [H^+]_{SWS}$$

In practical terms, the three seawater pH scales differ in their values by up to 0.12 pH units, differences that are much larger than the accuracy of pH measurements typically required, in particular, in relation to the ocean's carbonate system. Since it omits consideration of sulfate and fluoride ions, the free scale is significantly different from both the total and seawater scales. Because of the relative unimportance of the fluoride ion, the total and seawater scales differ only very slightly.

Living Systems

pH in living systems	
Compartment	pH
Gastric acid	1.5-3.5
Lysosomes	4.5
Human skin	4.7
Granules of chromaffin cells	5.5
Urine	6.0
Cytosol	7.2
Blood (natural pH)	7.34−7.45
Cerebrospinal fluid (CSF)	7.5
Mitochondrial matrix	7.5
Pancreas secretions	8.1

The pH of different cellular compartments, body fluids, and organs is usually tightly regulated in a process called acid-base homeostasis. The most common disorder in acid-base homeostasis is acidosis, which means an acid overload in the body, generally defined by pH falling below 7.35. Alkalosis is the opposite condition, with blood pH being excessively high.

The pH of blood is usually slightly basic with a value of pH 7.365. This value is often referred to as physiological pH in biology and medicine. Plaque can create a local acidic environment that can result in tooth decay by demineralization. Enzymes and other proteins have an optimum pH range and can become inactivated or denatured outside this range.

Calculations of pH

The calculation of the pH of a solution containing acids and/or bases is an example of a chemical speciation calculation, that is, a mathematical procedure for calculating the concentrations of all chemical species that are present in the solution. The complexity of the procedure depends on the nature of the solution. For strong acids and bases no calculations are necessary except in extreme situations. The pH of a solution containing a weak acid requires the solution of a quadratic equation. The pH of a solution containing a weak base may require the solution of a cubic equation. The general case requires the solution of a set of non-linear simultaneous equations.

A complicating factor is that water itself is a weak acid and a weak base. It dissociates according to the equilibrium,

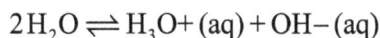

$$2H_2O \rightleftharpoons H_3O+ (aq) + OH- (aq)$$

with a dissociation constant, K_w defined as,

$$K_w = [H^+][OH-]$$

where [H+] stands for the concentration of the aqueous hydronium ion and [OH-] represents the concentration of the hydroxide ion. This equilibrium needs to be taken into account at high pH and when the solute concentration is extremely low.

where [H+] stands for the concentration of the aqueous hydronium ion and [OH-] represents the concentration of the hydroxide ion. This equilibrium needs to be taken into account at high pH and when the solute concentration is extremely low.

Strong Acids and Bases

Strong acids and bases are compounds that, for practical purposes, are completely dissociated in water. Under normal circumstances this means that the concentration of hydrogen ions in acidic solution can be taken to be equal to the concentration of the acid. The pH is then equal to minus the logarithm of the concentration value. Hydrochloric acid (HCl) is an example of a strong acid. The pH of a 0.01M solution of HCl is equal to $-\log_{10}(0.01)$, that is, pH = 2. Sodium hydroxide, NaOH, is an example of a strong base. The p[OH] value of a 0.01M solution of NaOH is equal to $-\log_{10}(0.01)$, that is, p[OH] = 2. From the definition of p[OH] above, this means that the pH is equal to about 12. For solutions of sodium hydroxide at higher concentrations the self-ionization equilibrium must be taken into account.

Self-ionization must also be considered when concentrations are extremely low. Consider, for example, a solution of hydrochloric acid at a concentration of $5 \times 10^{-8}M$. The simple procedure given above would suggest that it has a pH of 7.3. This is clearly wrong as an acid solution should have a pH of less than 7. Treating the system as a mixture of hydrochloric acid and the amphoteric substance water, a pH of 6.89 results.

Weak Acids and Bases

A weak acid or the conjugate acid of a weak base can be treated using the same formalism.

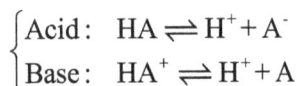

$$\begin{cases} \text{Acid:} & HA \rightleftharpoons H^+ + A^- \\ \text{Base:} & HA^+ \rightleftharpoons H^+ + A \end{cases}$$

First, an acid dissociation constant is defined as follows. Electrical charges are omitted from subsequent equations for the sake of generality

$$K_a = \frac{[H][A]}{[HA]}$$

and its value is assumed to have been determined by experiment. This being so, there are three unknown concentrations, [HA], [H$^+$] and [A$^-$] to determine by calculation. Two additional equations are needed. One way to provide them is to apply the law of mass conservation in terms of the two "reagents" H and A.

$$C_A = [A] + [HA]$$
$$C_H = [H] + [HA]$$

C stands for analytical concentration. In some texts, one mass balance equation is replaced by an equation of charge balance. This is satisfactory for simple cases like this one, but is more difficult to apply to more complicated cases as those below. Together with the equation defining K_a, there are now three equations in three unknowns. When an acid is dissolved in water $C_A = C_H = C_a$, the concentration of the acid, so [A] = [H]. After some further algebraic manipulation an equation in the hydrogen ion concentration may be obtained.

$$[H]^2 + K_a[H] - K_a C_a = 0$$

Solution of this quadratic equation gives the hydrogen ion concentration and hence p[H] or, more loosely, pH. This procedure is illustrated in an ICE table which can also be used to calculate the pH when some additional (strong) acid or alkaline has been added to the system, that is, when $C_A \neq C_H$.

For example, what is the pH of a 0.01M solution of benzoic acid, pK_a = 4.19?

- Step 1: $K_a = 10^{-4.19} = 6.46 \times 10^{-5}$

- Step 2: Set up the quadratic equation. $[H]^2 + 6.46 \times 10^{-5}[H] - 6.46 \times 10^{-7} = 0$

- Step 3: Solve the quadratic equation. $[H+] = 7.74 \times 10^{-4};\quad pH = 3.11$

For alkaline solutions an additional term is added to the mass-balance equation for hydrogen. Since addition of hydroxide reduces the hydrogen ion concentration, and the hydroxide ion concentration is constrained by the self-ionization equilibrium to be equal to $\dfrac{K_w}{[H+]}$

$$C_H = \frac{[H] + [HA] - K_w}{[H]}$$

In this case the resulting equation in [H] is a cubic equation.

General Method

Some systems, such as with polyprotic acids, are amenable to spreadsheet calculations. With three or more reagents or when many complexes are formed with general formulae such as $A_pB_qH_r$, the following general method can be used to calculate the pH of a solution. For example, with three reagents, each equilibrium is characterized by an equilibrium constant, β.

$$[A_p B_q H_r] = \beta_{pqr}[A]^p[B]^q[H]^r$$

Next, write down the mass-balance equations for each reagent:

$$C_A = [A] + \Sigma p\beta_{pqr}[A]^p[B]^q[H]^r$$

$$C_B = [B] + \Sigma q\beta_{pqr}[A]^p[B]^q[H]^r$$

$$C_H = [H] + \Sigma r\beta_{pqr}[A]^p[B]^q[H]^r - K_w[H]^{-1}$$

Note that there are no approximations involved in these equations, except that each stability constant is defined as a quotient of concentrations, not activities. Much more complicated expressions are required if activities are to be used.

There are 3 non-linear simultaneous equations in the three unknowns, [A], [B] and [H]. Because the equations are non-linear, and because concentrations may range over many powers of 10, the solution of these equations is not straightforward. However, many computer programs are available which can be used to perform these calculations. There may be more than three reagents. The calculation of hydrogen ion concentrations, using this formalism, is a key element in the determination of equilibrium constants by potentiometric titration.

Buffer Solution

A buffer is a solution that can resist pH change upon the addition of an acidic or basic components. It is able to neutralize small amounts of added acid or base, thus maintaining the pH of the solution relatively stable. This is important for processes and/or reactions which require specific and stable pH ranges. Buffer solutions have a working pH range and capacity which dictate how much acid/base can be neutralized before pH changes, and the amount by which it will change.

Composition of Buffer

To effectively maintain a pH range, a buffer must consist of a weak conjugate acid-base pair, meaning either a. a weak acid and its conjugate base, or b. a weak base and its conjugate acid. The use of one or the other will simply depend upon the desired pH when preparing the buffer. For example, the following could function as buffers when together in solution:

- Acetic acid (weak organic acid w/ formula CH_3COOH) and a salt containing its conjugate base, the acetate anion (CH_3COO^-), such as sodium acetate (CH_3COONa).

- Pyridine (weak base w/ formula C_5H_5N) and a salt containing its conjugate acid, the pyridinium cation ($C_5H_5NH^+$), such as Pyridinium Chloride.

- Ammonia (weak base w/ formula NH_3) and a salt containing its conjugate acid, the ammonium cation, such as Ammonium Hydroxide (NH_4OH).

Working of Buffer

A buffer is able to resist pH change because the two components (conjugate acid and conjugate base) are both present in appreciable amounts at equilibrium and are able to neutralize small amounts of other acids and bases (in the form of H_3O^+ and OH^-) when the are added to the solution. To clarify this effect, we can consider the simple example of a Hydrofluoric Acid (HF) and Sodium Fluoride (NaF) buffer. Hydrofluoric acid is a weak acid due to the strong attraction between the relatively small F^- ion and solvated protons (H_3O^+), which does not allow it to dissociate completely in water. Therefore, if we obtain HF in an aqueous solution, we establish the following equilibrium with only slight dissociation ($K_a(HF) = 6.6 \times 10^{-4}$, strongly favors reactants):

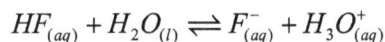

$$HF_{(aq)} + H_2O_{(l)} \rightleftharpoons F^-_{(aq)} + H_3O^+_{(aq)}$$

We can then add and dissolve sodium fluoride into the solution and mix the two until we reach the desired volume and pH at which we want to buffer. When Sodium Fluoride dissolves in water, the reaction goes to completion, thus we obtain:

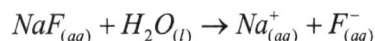

$$NaF_{(aq)} + H_2O_{(l)} \rightarrow Na^+_{(aq)} + F^-_{(aq)}$$

Since Na+ is the conjugate of a strong base, it will have no effect on the pH or reactivity of the buffer. The addition of NaF to the solution will, however, increase the concentration of F- in the buffer solution, and, consequently, by Le Châtelier's Principle, lead to slightly less dissociation of the HF in the previous equilibrium, as well. The presence of significant amounts of both the conjugate acid, HF, and the conjugate base, F-, allows the solution to function as a buffer. This buffering action can be seen in the titration curve of a buffer solution.

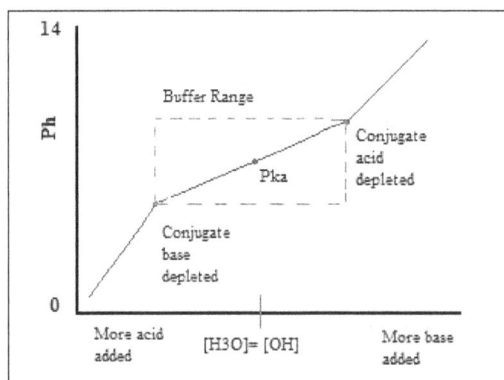

As we can see, over the working range of the buffer. pH changes very little with the addition of acid or base. Once the buffering capacity is exceeded the rate of pH change quickly jumps. This occurs because the conjugate acid or base has been depleted through neutralization. This principle implies that a larger amount of conjugate acid or base will have a greater buffering capacity.

If acid were added:

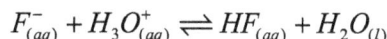

$$F^-_{(aq)} + H_3O^+_{(aq)} \rightleftharpoons HF_{(aq)} + H_2O_{(l)}$$

In this reaction, the conjugate base, F⁻, will neutralize the added acid, H_3O^+, and this reaction goes to completion, because the reaction of F⁻ with H_3O^+ has an equilibrium constant much greater than one. (In fact, the equilibrium constant the reaction as written is just the inverse of the K_a for HF: $1/K_a(HF) = 1/(6.6 \times 10^{-4}) = 1.5 \times 10^{+3}$.) So long as there is more F⁻ than H_3O^+, almost all of the H_3O^+ will be consumed and the equilibrium will shift to the right, slightly increasing the concentration of HF and slightly decreasing the concentration of F⁻, but resulting in hardly any change in the amount of H_3O^+ present once equilibrium is re-established.

If base were added:

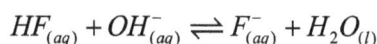

$$HF_{(aq)} + OH^-_{(aq)} \rightleftharpoons F^-_{(aq)} + H_2O_{(l)}$$

In this reaction, the conjugate acid, HF, will neutralize added amounts of base, OH⁻, and the equilibrium will again shift to the right, slightly increasing the concentration of F⁻ in the solution and decreasing the amount of HF slightly. Again, since most of the OH⁻ is neutralized, little pH change will occur.

These two reactions can continue to alternate back and forth with little pH change.

Selecting Proper Components for Desired pH

Buffers function best when the pK_a of the conjugate weak acid used is close to the desired working range of the buffer. This turns out to be the case when the concentrations of the conjugate acid and conjugate base are approximately equal (within about a factor of 10). For example, we know the K_a for hydroflouric acid is 6.6×10^{-4} so its $pK_a = -\log(6.6 \times 10^{-4}) = 3.18$. So, a hydrofluoric acid buffer would work best in a buffer range of around pH = 3.18.

For the weak base ammonia (NH_3), the value of K_b is 1.8×10^{-5}, implying that the K_a for the dissociation of its conjugate acid, NH_4^+, is $K_w/K_b = 10^{-14}/1.8 \times 10^{-5} = 5.6 \times 10^{-10}$. Thus, the pK_a for NH_4^+ = 9.25, so buffers using NH_4^+/NH_3 will work best around a pH of 9.25. (It's always the pK_a of the conjugate acid that determines the approximate pH for a buffer system, though this is dependent on the pK_b of the conjugate base, obviously.)

When the desired pH of a buffer solution is near the pK_a of the conjugate acid being used (i.e., when the amounts of conjugate acid and conjugate base in solution are within about a factor of 10 of each other), the Henderson-Hasselbalch equation can be applied as a simple approximation of the solution pH.

Adding Strong Acids or Bases to Buffer Solutions

Now that we have this nice F⁻/HF buffer, let's see what happens when we add strong acid or base to it. Recall that the amount of F⁻ in the solution is 0.66M x 0.1 L = 0.066 moles and the amount of HF is 1.0 M x 0.1L = 0.10 moles. Let's double check the pH using the Henderson-Hasselbalch Approximation, but using moles instead of concentrations:

$$pH = pK_a + \log(Base/Acid) = 3.18 + \log(0.066 \text{ moles F}^-/0.10 \text{ moles HF}) = 3.00$$

Good. Now let's see what happens when we add a small amount of strong acid, such as HCl. When we put HCl into water, it completely dissociates into H_3O^+ and Cl^-. The Cl^- is the conjugate base of a strong acid so is inert and doesn't affect pH, and we can just ignore it. However, the H_3O^+ can affect pH and it can also react with our buffer components. In fact, we already discussed what happens. The equation is:

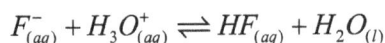

$$F^-_{(aq)} + H_3O^+_{(aq)} \rightleftharpoons HF_{(aq)} + H_2O_{(l)}$$

For every mole of H_3O^+ added, an equivalent amount of the conjugate base (in this case, F^-) will also react, and the equilibrium constant for the reaction is large, so the reaction will continue until one or the other is essentially used up. If the F^- is used up before reacting away all of the H_3O^+, then the remaining H_3O^+ will affect the pH directly. In this case, the capacity of the buffer will have been exceeded - a situation one tries to avoid. However, for our example, let's say that the amount of added H_3O^+ is smaller than the amount of F^- present, so our buffer capacity is NOT exceeded. For the purposes of this example, we'll let the added H_3O^+ be equal to 0.01 moles (from 0.01 moles of HCl). Now, if we add 0.01 moles of HCl to 100 mL of pure water, we would expect the pH of the resulting solution to be 1.00 (0.01 moles/0.10 L = 0.1 M; pH = -log(0.1) = 1.0).

However, we are adding the H_3O^+ to a solution that has F^- in it, so the H_3O^+ will all be consumed by reaction with F^-. In the process, the 0.066 moles of F^- is reduced:

0.066 initial moles F^- - 0.010 moles reacted with H_3O^+ = 0.056 moles F^- remaining

Also during this process, more HF is formed by the reaction:

0.10 initial moles HF + 0.010 moles from reaction of F^- with H_3O^+ = 0.11 moles HF after reaction

Plugging these new values into Henderson-Hasselbalch gives:

pH = pK_a + log (base/acid) = 3.18 + log (0.056 moles F^-/0.11 moles HF) = 2.89

Thus, our buffer did what it should - it resisted the change in pH, dropping only from 3.00 to 2.89 with the addition of 0.01 moles of strong acid.

pOH

pOH is a measure of hydroxide ion (OH-) concentration. It is used to express the alkalinity of a solution.

Aqueous solutions at 25°C with pOH less than 7 are alkaline, pOH greater than 7 are acidic and pOH equal to 7 are neutral.

pOH is calculated based on pH or hydrogen ion concentration ($[H^+]$). Hydroxide ion concentration and hydrogen ion concentration are related:

$[OH^-] = K_w / [H^+]$

K_w is the self-ionization constant of water. Taking the logarithm of both sides of the equation:

$$pOH = pK_w - pH$$

An approximation is that:

$$pOH = 14 - pH$$

While the approximation works well in many settings, there are exceptions when the pK_w value should be used instead.

References

- Introduction-to-acids, acids-bases-and-salts, chemistry, guides: toppr.com, Retrieved 30 March, 2019

- Johll, matthew e. (2009). Investigating chemistry: a forensic science perspective (2nd ed.). New york: w. H. Freeman and co. Isbn 1429209895. Oclc 392223218

- "what is triacidic? Definition of triacidic (science dictionary)". Science dictionary. 14 september 2013. Retrieved 14 march 2019

- Definition-of-poh-in-chemistry-605893: thoughtco.com, Retrieved 5 August, 2019

- "electrophile - nucleophile - basicity - acidity - ph scale". City collegiate. Archived from the original on 30 June 2010. Retrieved 20 June 2016

- Acid-base-reaction, science: britannica.com, Retrieved 2 July, 2019

- Nørby, jens (2000). "the origin and the meaning of the little p in ph". Trends in biochemical sciences. 25 (1): 36–37. Doi:10.1016/s0968-0004(99)01517-0. Pmid 10637613

6
Chemical Mixtures and Solutions

Chemical mixture is a material that is made up of two or more different substances which are physically combined. The mixture could be homogeneous or heterogeneous in nature. Solutions are a type of homogenous mixture which consists of a solute that is dissolved in another sub-stance, known as a solvent. The diverse types of chemical mixtures and solutions have been thoroughly discussed in this chapter.

Mixture

In chemistry, a mixture forms when two or more substances are combined such that each substance retains its own chemical identity. Chemical bonds between the components are neither broken nor formed. Note that even though the chemical properties of the components haven't changed, a mixture may exhibit new physical properties, like boiling point and melting point. For example, mixing together water and alcohol produces a mixture that has a higher boiling point and lower melting point than alcohol (lower boiling point and higher boiling point than water).

Examples of Mixtures

- Flour and sugar may be combined to form a mixture.

- Sugar and water form a mixture.

- Marbles and salt may be combined to form a mixture.

- Smoke is a mixture of solid particles and gases.

Types of Mixtures

Two broad categories of mixtures are heterogeneous and homogeneous mixtures. Heterogeneous mixtures are not uniform throughout the composition (e.g. gravel), while homogeneous mixtures have the same phase and composition, no matter where you sample them (e.g., air). The distinction between heterogeneous and homogeneous mixtures is a matter of magnification or scale. For example, even air can appear to be heterogeneous if your sample only contains a few molecules, while a bag of mixed vegetables may appear homogeneous if your sample is an entire truckload full of them. Also note, even if a sample consists of a single element, it may form a heterogeneous mixture. One example would be a mixture of pencil lead and diamonds (both carbon). Another example could be a mixture of gold powder and nuggets.

Besides being classified as heterogeneous or homogeneous, mixtures may also be described according to the particle size of the components:

- Solution - A chemical solution contains very small particle sizes (less than 1 nanometer in diameter). A solution is physically stable and the components cannot be separated by decanting or centrifuging the sample. Examples of solutions include air (gas), dissolved oxygen in water (liquid), and mercury in gold amalgam (solid), opal (solid), and gelatin (solid).

- Colloid - A colloidal solution appears homogeneous to the naked eye, but particles are apparent under microscope magnification. Particle sizes range from 1 nanometer to 1 micrometer. Like solutions, colloids are physically stable. They exhibit the Tyndall effect. Colloid components can't be separated using decantation, but may be isolated by centrifugation. Examples of colloids include hair spray (gas), smoke (gas), whipped cream (liquid foam), blood (liquid).

- Suspension - Particles in a suspension are often large enough that the mixture appears heterogeneous. Stabilizing agents are required to keep the particles from separating. Like colloids, suspensions exhibit the Tyndall effect. Suspensions may be separated using either decantation or centrifugation. Examples of suspensions include dust in air (solid in gas), vinaigrette (liquid in liquid), mud (solid in liquid), sand (solids blended together), and granite (blended solids).

Homogeneous Mixture

A homogeneous mixture is simply any mixture that is uniform in composition throughout. Below, you will find many examples of homogeneous mixtures.

Everyday Homogeneous Mixtures

Here are some homogeneous mixtures:

- Water itself is an example of a homogeneous mixture. It often contains dissolved minerals and gases, but these are dissolved throughout the water. Tap water and rain water are both homogeneous, even though they may have different levels of dissolved minerals and gases.

- A bottle of alcohol is a man-made homogeneous mixture, from a fine Italian wine to a glass of Scotch whisky.

- In the human body, blood plasma is an example of a homogeneous mixture. This is a colorless fluid that holds the blood cells in suspension. It makes up a little more than half of the volume of human blood.

- A cup of coffee is considered a homogeneous mixture. Does that change when sugar is dissolved into it? As a matter of fact, if the sugar is completely dissolved, the coffee is still considered homogeneous, since the mixture of coffee and sugar is uniform throughout the cup.

- Mouthwash is a homogeneous mixture example. Mouthwashes typically contain a percentage of alcohol along with a variety of different chemicals aimed at keeping the teeth and gums clean and the breath fresh.

- Laundry detergent is another example of a homogeneous mixture of various soaps and chemicals that keep clothes clean.

- A bottle of vinegar is a man made homogeneous mixture. Many people use it to cook, and it's also popular as a cleaning product.

- The air that you breathe is a homogeneous mixture of oxygen, nitrogen, argon, and carbon dioxide, along with other elements in smaller amounts. Because each layer of the Earth's atmosphere has a different density, each layer of air is its own homogeneous mixture.

- That perfume or cologne you use to smell good is a perfectly homogeneous mixture of chemicals and dyes.

- Many acids and solutions are homogeneous mixtures - for example, a solution of diluted hydrochloric acid.

- Gunpowder, which is used in a variety of explosives, is actually a homogeneous mixture of elements like sulfur, charcoal, and saltpeter (also known as potassium nitrate).

- Many beverages are examples of homogeneous mixtures, from the orange juice you drink in the morning to the glass of water you drink in the afternoon and the cup of tea you have just before bed.

- Put some soap into a glass of warm water, cover it, and shake it up - the mixture of soapy water will in fact be homogeneous in composition.

- An alloy is a metal comprised of two pure metals. Alloys such as steel and bronze are homogeneous mixtures of two metals.

- The bottle of vegetable oil so often used in cooking is a homogeneous mixture.

- Jello gelatin is a colloid, a special type of homogeneous mixture that has particles microscopically dispersed through the substance. Other examples of this type of homogeneous mixture are a cloud of dust, a glass of milk, and honey made by bees.

Heterogeneous Mixture

A heterogeneous mixture is simply any mixture that is not uniform in composition - it's a non-uniform mixture of smaller constituent parts. Using various means, the parts in the mixture can be separated from one another.

Everyday Heterogeneous Mixtures

Here are some examples of very common heterogeneous mixtures:

- A bowl of Fruit Loops cereal is a heterogeneous mixture because it has cereal bits of many colors floating around in milk.

- A bottle of balsamic vinaigrette salad dressing is a mixture that is heterogeneous, and has to be shaken up to make the mixture appear and taste more combined.

- Sand shaken up in a bottle of water is a heterogeneous mixture of sand particles floating around which will eventually settle to the bottom of the bottle, making it look a lot less like a mixture.

- Rocks in the sand at the beach are a heterogeneous mixture - all different shapes, sizes and colors - just thrown together at random.

- Smog is a heterogeneous mixture of various particles suspended in the air. The dirty particles that make up the smog can be removed from the air and breathed into the lungs, making smog quite a problematic heterogeneous mixture.

- Mixed nuts at a party is a type of heterogeneous mixture that can be separated simply by pouring them onto a table and sorting them into separate piles, each of the same nut.

- A puddle of mud counts as a heterogeneous mixture - dirt, leaves, grasses, animal byproducts - all blended together in water.

- Pizza is a heterogeneous mixture of dough, sauce, cheese, and other toppings.

- A bowl of oatmeal with raisins is a heterogeneous mixture.

- A salad with lettuce, cheese, seeds, tomatoes, broccoli, and other vegetables is an example of a heterogeneous mixture.

- The Pacific Ocean is a huge example of a heterogeneous mixture, with all type of plants and animals floating around haphazardly, suspended in the salt water.

- A bowl of candy like M&Ms, Skittles, or jelly beans is a heterogeneous mixture, with a non-uniform variety of colors and flavors represented in a single bowl.

- Soil is an example of a heterogeneous mixture that combines many different elements which are not uniform.

- Hollandaise sauce on your eggs may not look like a heterogeneous mixture - but it is! It's a special type of mixture called an emulsion that can break apart into its constituent pieces (butter, egg yolks, and lemon juice).

- An atom is a heterogeneous mixture of various parts such as protons, electrons, and neutrons. As demonstrated with nuclear weapons, atoms can actually be split apart using certain methods.

- Sand is a heterogeneous mixture of rock, shells, metals, and other elements, which can be separated from each other by methods like sifting.

- Vinegar and oil are often mixed together as a condiment, but the mixture itself is heterogeneous. They may stay together for a while, but they're sure to be broken apart after a while.

Saturated Solution

The term saturated solution is used in chemistry to define a solution in which no more solute can be dissolved in the solvent. It is understood that saturation of the solution has been achieved when any additional substance that is added results in a solid precipitate or is let off as a gas.

Understanding Saturated Solutions

There are many different factors that can affect whether something is a saturated solution. For example, saturation is affected by:

- The solution's temperature.

- The solution's pressure.

- Chemical makeup of substances involved.

Ways to make a saturated solution include:

- Add solute to liquid until dissolving stops.

- Evaporate a solvent from a solution until the solute begins to crystallize or precipitate.

- Add seed crystals to a solution that is supersaturated.

Everyday Examples of Saturated Solutions

- Carbonated water is saturated with carbon, hence it gives off carbon through bubbles.

- Adding sugar to water until it no longer dissolves creates a saturated solution.

- Continuing to dissolve salt in water until it will no longer dissolve creates a saturated solution.

- The Earth's soil is saturated with nitrogen.

- Mixing powdered soap into water until it will not dissolve creates a saturated solution.

- In beer or sparkling juices there is a saturation of carbon dioxide that is let off as a gas.

- Coffee powder added to water can create a saturated solution.

- Salt added to vinegar can create a saturated solution when the salt no longer dissolves.

- Chocolate powder added to milk can create saturation at the point that no more powder can be added.

- Sugar dissolved into vinegar until it will no longer do so creates a saturate solution.

- Water can be saturated with juice powder to create a beverage.

- Milk can be saturated with flour at which point no more flour can be added to the milk.

- Melted butter can be saturated with salt when the salt will no longer dissolve.

- Bathing salts can saturate water when there is no more ability to dissolve them.

- Sugar can be added to milk to the point of saturation.

- Processed tea powders can be added to water to saturate the water.

- Protein powder could be used to create a saturated solution with milk, tea, or water.

- Laxative powders could saturate juice or water with which they are mixed.

- Cocoa powder could be mixed into water to the point of saturation.

- Sugar could be mixed into tea to the point that the tea is saturated.

- Coffee can be saturated with sugar when no more will mix in to the coffee.

Things that are insoluble in water cannot create the saturated solutions. For example, pepper and sand can not be dissolved in water and therefore cannot create a saturated solution.

However, now you have seen some examples of different ways that a saturated solution can be created and different kinds of saturated solutions.

Unsaturated Solution

An unsaturated solution is a chemical solution in which the solute concentration is lower than its equilibrium solubility. All of the solute dissolves in the solvent.

When a solute (often a solid) is added to a solvent (often a liquid), two processes occur simultaneous. Dissolution is the dissolving of the solute into the solvent. Crystallization is is the opposite process, where the reaction deposits solute. In an unsaturated solution, the rate of dissolution is much greater than the rate of crystallization.

Examples of Unsaturated Solutions

- Adding a spoonful of sugar to a cup of hot coffee produces an unsaturated sugar solution.

- Vinegar is an unsaturated solution of acetic acid in water.

- Mist is an unsaturated (but close to saturated) solution of water vapor in air.

- 0.01 M HCl is an unsaturated solution of hydrochloric acid in water.

Types of Saturation

There are three levels of saturation in a solution:

- In an unsaturated solution there is less solute than the amount that can dissolve, so it all goes into solution. No undissolved material remains.

- A saturated solution contains more solute per volume of solvent than an unsaturated solution. The solute has dissolved until no more can, leaving undissolved matter in the solution. Usually the undissolved material is more dense than the solution and sinks to the bottom of the container.

- In a supersaturated solution, there is more dissolved solute than in a saturated solution. The solute can easily fall out of solution by crystallization or precipitation. Special conditions may be needed to supersaturate a solution. It helps to heat a solution to increase solubility so more solute can be added. A container free of scratches also helps keep solute from falling out of solution. If any undissolved material remains in a supersaturated solution, it can act as nucleation sites for crystal growth.

Components of a Solution

Solute

A solute is a substance that can be dissolved by a solvent to create a solution. A solute can come in many forms. It can be gas, liquid, or solid. The solvent, or substance that dissolves the solute, breaks the solute apart and distributes the solute molecules equally. This creates a *homogenous mixture*, or solution that is equal throughout.

Solutes in solution are measured by their *concentration*. The concentration of a solute is the amount of solute divided by the total volume of solution. A solvent can dilute various amounts of solute, depending on how strong of a solvent is used and how easily the solute molecules come apart. This property of solutes to dissolve in a solvent is known as *solubility*.

Examples of Solute

Salt in Water

When you dump a spoon full of salt into a glass of water, you are creating a solution. The solute is the salt, or NaCl. The solvent is water, or H_2O. The water molecules are negatively charged on the oxygen atoms and positively charged on the hydrogen atoms. Salt is an *ionic compound*, which consists of two ions: Na^+ and Cl^-. The negative oxygen atoms attract the positive sodium (Na^+), and the positive hydrogen atoms attract the negative chlorine atoms (Cl^-). The attraction between the different molecules pulls the solute apart at a molecular level, and suspends it evenly throughout the water.

An important factor in how fast the solute will dissolve is the surface area of solute exposed. If coarse salt is used, less surface area is exposed and it will take longer for the same amount of salt to dissolve. A finer salt allows many more ions to be exposed to water, and the solute gets diffused through the water faster. Eventually the salt can no longer be seen on the bottom of the glass because it is evenly distributed throughout the glass.

A similar process happens with sugar, but the sugar molecules are not the same as salt molecules. Instead of being an ionic compound, the sugar molecules are slightly polar. The molecule of sugar has many OH groups, which create natural dipoles. These positive and negative areas interact with the positive and negative areas of the water molecules, and the solute molecule are torn apart. Just as salt is diffused across a solution, sugar can also be evenly distributed in a cell. This is important for many cellular functions, such as producing energy and larger molecules. Other times, cells must actively transport certain molecules out of the cytosol, to avoid upsetting the pH balance.

Oxygen in Seawater

An example of a gaseous solute is oxygen. Every fish in the ocean, from the strange creatures in the deepest parts of the ocean to the common coral-dwelling fish that scuba divers love, rely on oxygen dissolved in the water to live. The oxygen, which exists as O_2, is a polar molecule. As such, the polar water molecules have a natural tendency to attract the oxygen. As the waves mix air into the ocean and the surface of the ocean and atmosphere interact, oxygen is dissolved into the water. The process of diffusion carries the oxygen through the water column, delivering oxygen to organisms throughout the ocean.

In some situations the organisms in the ocean can use the oxygen in the water faster than it can be diffused into the water. This can happen when excess nutrient runoff from humans runs into the ocean. The nutrients, which are another solute in water, allow huge *algal blooms* to grow. These blooms contain far too many algae. The algae in the lower layers start to die, and bacteria start to consume them. Between the algae and the bacteria, all of the oxygen gets used up. This creates a dead zone in the water column. If fish start to swim through this column, they could suffocate from lack of oxygen.

Protons in the Cytosol

Organisms of all kinds must regulate the amount of solutes in their cells, to maintain proper cell functions. The acidity of cells is based in part on the number of hydrogen ions (H^+), or protons, found in the solution of cytosol. The protons are attracted to the oxygen atoms of water, because they are electronegative. The protons as a solute serve a very important function in cells. While water is able to diffuse through a cellular membrane via osmosis, hydrogen atoms cannot breach the membrane. The concentration gradient creates a potential force that can be used to move other substances. This is known as *proton motive force* and is used to move a wide variety of substances through the cellular membrane.

Solvent

Solvent is a substance, ordinarily a liquid, in which other materials dissolve to form a solution. Polar solvents (e.g., water) favour formation of ions; nonpolar ones (e.g., hydrocarbons) do not.

Solvents may be predominantly acidic, predominantly basic, amphoteric (both), or aprotic (neither). Organic compounds used as solvents include aromatic compounds and other hydrocarbons, alcohols, esters, ethers, ketones, amines, and nitrated and halogenated hydrocarbons. Their chief uses are as media for chemical syntheses, as industrial cleaners, in extractive processes, in pharmaceuticals, in inks, and in paints, varnishes, and lacquers.

References

- Mixture-definition-chemistry-glossary-606374: thoughtco.com, Retrieved 25 February, 2019

- Examples-of-homogeneous-mixture: yourdictionary.com, Retrieved 26 July, 2019

- Examples-of-saturated-solution: yourdictionary.com, Retrieved 8 January, 2019

- Definition-of-unsaturated-solution-605936: thoughtco.com, Retrieved 13 May, 2019

- Solute: biologydictionary.net, Retrieved 6 January, 2019

- Solvent-chemistry, science: britannica.com, Retrieved 1 February, 2019

Permissions

Index

www.ingramcontent.com/pod-product-compliance
Lightning Source LLC
Chambersburg PA
CBHW082013190326
41458CB00010B/3177